Entity Authentication and Personal Privacy in Future Cellular Systems

RIVER PUBLISHERS SERIES IN STANDARDISATION

Volume 2

Standardisation is a book series addressing the pre-development related standards issues and standardized technologies already deployed. The focus of this series is also to examine the application domains of standardised technologies. This series will present works of foras and standardization bodies like IETF, 3GPP, IEEE, ARIB, TTA, CCSA, WiMAX, Bluetooth, ZigBee, etc.

Other than standards, this book series also presents technologies and concepts that have prevailed as *de-facto*.

Scope of this series also addresses prevailing applications which lead to regulatory and policy issues. This may also lead towards harmonization and standardization of activities across industries.

For a list of other books in this series, see final page.

Entity Authentication and Personal Privacy in Future Cellular Systems

Geir M. Køien

University of Agder, Norway

River Publishers

Aalborg

ISBN 978-87-92329-32-5 (hardback)

Published, sold and distributed by:
River Publishers
P.O. Box 1657
Algade 42
9000 Aalborg
Denmark

Tel.: +45369953197
www.riverpublishers.com

Contents

Preface

Access to Specifications

This book references many technical standards and recommendations for the 3GPP, the IETF and ITU. Fortunately, all these documents are freely available and so you may dig deeper yourself if you feel that the book does not provide enough details.

The technical specifications and technical reports of the 3GPP are all available at the 3GPP website (www.3gpp.org), with the exception of some of the cryptographic algorithms documents. The cryptographical algorithms are available too, but not always at the 3GPP servers, which are located in France. The French weapons export procedure (cryptographic methods are regarded as arms) works very slowly and some documents are therefore hosted by the GSM Association instead.

The interested reader may also want to look up IETF RFCs. These are conveniently found at http://rfc-editor.org/rfcsearch.html in both ascii text and Adobe PDF format. At http://tools.ietf.org/html/ the same RFCs can be found in HTML format.

The 3GPP TS/TR by series is found at http://www.3gpp.org/specification-numbering. From a security point of view the the most interesting series are the 33-series, the 35-series, the 43-series and the 55-series. The 3GPP TS/TR maintained by the 3GPP SA3 (Security) work group is found at http://www.3gpp.org/ftp/Specs/html-info/TSG-WG-S3.htm.

The crypto algorithm specifications are hosted by the *GSM Association* as a sub-page to www.gsmworld.com/our-work/programmes-and-initiatives/fraud-and-security/index.htm. The GSMA page contains additional material on practical mobile telephone security measures.

Recommendations from the ITU are also available at their website www.itu.int. You may locate the recommendations by series or you may use the website's internal search facility.

The 3GPP Releases

The 3GPP uses a system of parallel releases in which new functionality is added to new releases. The scheme is elaborated on in http://www.3gpp.org/releases.

3GPP Technical Standards

The 3GPP is not a standards body and it cannot formally ratify technical standards. However, the 3GPP organizational members are official standards bodies and they may approve and ratify the 3GPP documents for use within their respective regulatory and legal mandate. Nevertheless, 3GPP standards are normally ratified as-is by the organizational partners.

Acknowledgements

This book is partly based on my PhD thesis. I would therefore extend my gratitude to both my thesis supervisors Professor Vladimir A. Oleshchuk and Professor Ramjee Prasad for help and support with the PhD. I would also like to thank my previous employer, Telenor R&I, for having supported much of the work.

Furthermore, I would like to thank two of my former colleagues, Runar Langnes and Vidar Bjugan, who read and provided feedback on Chapter 2.

Additional thanks go to the men and women of the 3GPP SA3 (Security) work group. I have had the pleasure of being a delegate to SA3 for a number of years now and it has been and interesting experience to follow the progress of security standards over the years.

List of Figures

Milton Keynes UK
Ingram Content Group UK Ltd.
UKHW012036270824
447508UK00009B/185

List of Tables

Acronyms and Abbreviations

The following is an overview of the acronyms and abbreviations used in this book. TR 21.905 [1] contains a reasonable overview over 3GPP terms.

Term	Comment/Definition
3GPP	Third Generation Partner Project; project to develop 3G (UMTS) system. See www.3gpp.org
3GPP2	Third Generation Partner Project 2; project to develop CDMA2000 system. See www.3gpp2.org
AAA	Authentication, Authorization and Accounting
AES	Advanced Encryption Standard
AH	Authentication Header; part of the IPsec protocol suite
AK	Anonymity Key. 3GPP term. Used in conjunction with the SQN to mitigate tracking threat
AMF	Authentication Management Field; part of the AV
AN	Access Network. See also RAS
ANG	Access Network Gateway
AS	Access Stratum
ASME	Access Security Management Entity (LTE)
AP	Access Point
AuC	Authentication Centre
AUTN	Authentication Token; part of the AV. $AUTN := SQN \| AMF \| MAC\text{-}A$
AV	Authentication Vector; 3GPP term. $AV := RAND \| XRES \| CK \| IK \| AUTN$
AVISPA	Automated Validation of Internet Security Protocols and Applications. Name of IST project
BAN	Burrows, Abadi and Needham. Refers to "a Logic of Authentication"
BSC	Base station controller; GSM access network node
CA	CA – Channel over A-interface (UE–SN)
CB	CB – Channel over B-interface (SN–HE)
CBC	Cipher Block Chaining
CC	CC – Channel over A- and B-interface (UE–HE)
CDMA2000	A 3G cellular system. Developed by 3GPP2
CID	Context Identity; a medium-term reference identity to the UE. Allocated by the UE
CK	Confidentiality key; part of the AV
CL-AtSe	Constraint Logic Attack Searcher
CN	Core Network

CP   Control Plane
C-RNTI   Cell-Radio Network Temporary Identifier
CS   Circuit-Switched
CSP   Communicating Sequential Processes
DDoS/DoS   Distributed Denial-of-Service/Denial-of-Service
DH   Diffie–Hellman
PDCP   Packet Data Convergence Protocol
DTAP   Direct Transfer Application Part; protocol used in radio access
                (GSM/GPRS/UMTS)
DYI   Dolev-Yao Intruder
eKSI   Key Set Identifier (E-UTRAN specific)
eNB   Evolved NodeB
EAP   Extensible Authentication Protocol
EAP-AKA   EAP method for running (a slightly modified) UMTS AKA protocol
EAPOL   EAP over LAN; defined in IEEE 802.1X
ECC   Elliptic Curve Cryptography
EDGE   Enhanced Data rates for GSM Evolution; new radio technology in GSM
                frequency scheme
EEA   EPS Encryption Algorithm
EIA   EPS Integrity Algorithm
EPC   Evolved Packet Core (LTE)
EPS   Evolved Packet System (LTE)
EPS-AV   EPS Authentication Vector (LTE); also E-AV
ESP   Encapsulated Security Payload; part of the IPsec protocol suite
ETSI   European Telecommunications Standards Institute
E-UTRAN   Evolved UTRAN (LTE)
FDR   Failures-Divergences Refinement
GDDoS   Geographically Distributed Denial-of-Service
GERAN   GSM EDGE Radio Access Network
GGSN   Gateway GPRS Support Node
GPRS   General Packet Radio Service; originally developed as a packet switched
                extension to GSM
GSM   Global System for Mobile communications; 2G system, now maintained
                by 3GPP
GTP   GPRS Tunneling Protocol
GUTI   Globally Unique Temporary Identity (LTE)
GW   Gateway
HE   Home Environment (3GPP). Also: Home Entity
HEID   Home Entity Identity; public HE identity
HLPSL   High-Level Protocol Specification Language; language developed and used
                in the AVISPA project
HLR   Home Location Register
HN   Home Network
HOL   Higher Order Logic
HR   Home Register
HSS   Home Subscriber Server

IBE	Identity-Based Encryption
ICV	Integrity Check Value
IE	Information Element
IETF	Internet Engineering Task Force
IF	Intermediate Format
IK	Integrity key; part of the AV
IMEI	International Mobile Equipment Identity
IMS	IP Multimedia Subsystem
IMSI	International Mobile Subscriber Identity; (3GPP) Primary subscriber identity
IPsec	IP security (network layer). Defined by IETF
KASUMI	Name of standard UMTS crypto-primitives for the $f8$ and $f9$ functions
KDC	Key Distribution Center
KDF	Key Derivation Function
K_{ASME}	Master key for EPS security context
K_{eNB}	Root key for LTE-Uu security (based on K_{ASME})
K_{NASenc}	Session key for data confidentiality between ME and MME (NAS protocol)
K_{NASint}	Session key for data integrity between ME and MME (NAS protocol)
K_{RRCenc}	Session key for data confidentiality between ME and eNB (RRC protocol)
K_{RRCint}	Session key for data integrity between ME and eNB (RRC protocol)
K_{UPenc}	Session key for data confidentiality between ME and eNB (LTE-Uu)
LA	Location Area
LEA	Law Enforcement Agency
LI	Lawful Interception
LTC	Long-Term Context
LTE	Long-Term Evolution; a name for the evolved UMTS radio access system (Rel.8)
LTL	Linear Temporal Logic
MAC	Message Authentication Code/Medium Access Control
MAC-A	Message Authentication Code – Authentication; used as AV signature (part of the AV)
MAC-I	Message Authentication Code for Integrity – Terminology used in TS 36.323 [2]
MAP	Mobile Application Part
ME	Nobile Equipment
MILENAGE	Name of example set of authentication algorithms
MitM	Man-in-the-Middle
MME	Mobility Management Entity (LTE)
MS	Mobile Station/Mobile Subscriber; term used in GSM
MSC	Mobile Switching Centre
MSISDN	Mobile Subscriber ISDN Number
MT	Mobile Termination
MTC	Medium-Term Context
NAS	Non Access Stratum (LTE)
NCC	NH Chaining Counter (a term specific to TS 33.401)
NDS	Network Domain Security
NGN	Next Generation Network

NH	Next hop
NMT	Nordic Mobile Telephony; NMT was a 1G analogue cellular system
NodeB	UTRAN access point (base station)
OFM	Output-Feedback Mode. Also sometimes known as OFB
OFMC	On-the-Fly Model-Checker
PCI	Physical Cell Identity
PDCP	Packet Data Convergence Protocol
PE	Prudent Engineering (design principles)
PE3WAKA	Privacy Enhanced 3-Way Authentication and Key Agreement; name of protocol family
PET	Privacy Enhancing Technology
PIN	Personal Identification Number
PKI	Public Key Infrastructure
PLMN	Public Land Mobile Network
PS	Packet-Switched
PUK	PIN Unblocking Key
R-UIM	Removable-UIM; 3GPP2 term. Smart card base UIM (similar to USIM on UICC in 3GPP terms)
RAN	Radio Access Network
RAS	Radio Access Server
RES	Response; corresponding to XRES in the AV. Also used generically
RNC	Radio Network Controller; UTRAN node
RRC	Radio Resource Control
S2PLIP	Secure Two-Party Location Inclusion Protocol
SAE	System Architecture Evolution; a name for the evolved UMTS core network (Rel.8)
SGSN	Serving GPRS Support Node
SID	Session Identity; a short-term alias identity for the UE. Assigned by the SN
SM	Secure Module
SMC	Secure Multi-part Computation/Security Mode Command
SN	Serving Network
$SNID$	Serving Network Identity. Public SN identity
SNOW-3G	Name of (second) standard UMTS crypto-primitives for the $f8$ and $f9$ functions
SNS	Serving Network Server
SPAN	Security Protocol ANimator
SQN	Sequence Number; part of the AUTN part of the AV
SS7	signalling System No.7; ITU-T protocol stack for system signalling
STC	Short-Term Context
TDMA	Time Division Multiple Access; access scheme used in GSM
TISPAN	Telecommunication and Internet converged Services and Protocols for Advanced Networking
TLA	Temporal Logic of Actions
TMSI	Temporary Mobile Subscriber Identity
TAU	Tracking Area Update (NAS protocol message name)
TVP	Time-Variant Parameter

UD	User Device
UE	User Equipment; term used in 3GPP specifications
UEA	UMTS Encryption Algorithm; algorithm identifier for the $f8$ function
UEID	User Entity Identity; the permanent UE identity
UIA	UMTS Integrity Algorithm. Algorithm identifier for the $f9$ function
UICC	A physically secure device, an IC card (or "smart card")
UIM	User Identity Module; 3GPP2 term
UMTS	Universal Mobile Telecommunications System; a 3G cellular system developed by 3GPP
UP	User Plane
USIM	Universal Subscriber Identity Module; security module for UMTS subscriptions
UTRAN	Universal Terrestrial Radio Access Network; UMTS access network
VASP	Value Added Service Provider
VLR	Visitor Location Register
WEP	Wired Equivalent Privacy; old deprecated security scheme for IEEE 802.11 WLAN. Not secure
WLAN	Wireless Local Area Network; often associated with the IEEE 802.11 standards
WPA	WiFi Protected Access; profiled use of IEEE 802.11i methods. Two main variants (WPA/WPA-2)
XRES	Expected response; part of the AV. See also RES

1

Introduction

> Those that can give up essential liberty to obtain a little safety
> deserve neither liberty nor safety.
> — *Benjamin Franklin*

1.1 Background

1.1.1 What This Book Is About

In the first quarter of 2009, there were more than 4 billion subscribers to cellular phone services in the world.[1] With this in mind it should be clear that use of mobile communication has already become both pervasive and ubiquitous. A truly global commodity.

This book aims at explaining and examining access security as it is found in mobile/cellular systems. In particular, we shall go through how access security and personal privacy is handled in the 3GPP systems, which includes both the 2G systems GSM/GPRS, the 3G system UMTS and the 4G system LTE.[2] The first part of the book deals exclusively with presenting access security, and in particular the authentication and key agreement procedures, in the 3GPP system.

The second part will serve to highlight both the weaknesses and the strengths of the 3GPP systems and it will also show that more could have been done. Thus in the second part we go on to examine what is missing from the current cellular access security architectures. Some of the shortcomings found in UMTS have been partially addressed in LTE, but the heavy burden of backwards compatibility has meant that many issues could not be resolved

---

[1] GSM Association (http://www.gsmworld.com/).

[2] Formally, LTE is still not 4G. This will be the role of the so-called LTE Advanced. However, the differences are mostly concerned with radio access technology capabilities and the security architecture will be the same for LTE and LTE Advanced.

easily. We shall see that one can provide privacy enhanced entity authentication and one can avoid delegated authentication control. We will also look into the design of authentication protocols and even look at the role of formal verification in the design of security protocols.

1.1.2 Justification

Why Improve on Cellular Access Security?
The current cellular systems, epitomized by the UMTS system, have a fairly long history. The basic cellular architecture was already in place in first generation systems like the Nordic Mobile Telephony (NMT) system launched in 1981 [3]. This system can be traced back to the Nordic national tele-directors meeting in 1969 in Kabelvåg, Norway. At the same time an agreement was signed to permit roaming[3] [4]. The trust scenarios were very different at the time and the available technology could not really provide much security (the transmission was analogue and would have been difficult to protect anyway).

Later, with the arrival of 2G systems like GSM, the need for security was more apparent and the technology had matured. However, the generic trust relations were the same or fairly similar as for the 1G systems. So the GSM security architecture [5] was built around the delegated trust model from the 1G systems. Sadly, the 3G system (UMTS, CDMA2000) retained much of the delegated trust model, in which the home network operator delegates the responsibility for carrying out authentication and key agreement to the serving network. While the trust model was permissible for the 1G system, it was questionable for the 2G systems and, frankly, it was quite inadequate for the 3G systems [6, 7]. Not to speak of the 4G LTE system.

So there is a need for developing a more sophisticated model of the trust relations and to properly take into account that there should be three active participants in the authentication and key agreement procedures.

For cellular systems the efficiency and speed of the security protocol is critical. The *registration* procedure, the *handover* procedure and the *connection setup* procedures are all real-time procedures. The deadline with respect to the physical radio environment is tight, but not necessarily fixed. This means that any procedure that is executed in conjunction with these main mobility procedures must be fast.

---

[3] This was probably the first roaming agreement ever. It was all kept within one A4 page.

Why Provide Privacy Enhancements?

The current UMTS and LTE security architectures already contain a set of subscriber privacy requirements. These acknowledge the need for identity privacy and location privacy for the subscribers. The requirements do address some important personal privacy aspects, but the requirements are incomplete [8].

The right to privacy is not limited or restricted to any technical mechanism or to any contemporary technology. It has a deeper and more philosophical side to it. These aspects are not entirely new, but are becoming more and more important in an all-digital world where the rights to personal privacy can so easily be violated.

In the paper "The Right to Privacy" [9] the authors give a broad and compelling range of arguments for one's right to personal privacy. Amongst the diverse set of privacy rights discussed is the right "to be left alone", which is indeed close in spirit to the location and identity privacy described in this book. The authors continue that "... the individual is entitled to decide whether that which is his shall be given to the public." The arguments go beyond aspects such as copyrights or property rights and focus on privacy as a generic right not to disclose personal information without due consent.

The authors write from a perspective of generic legal rights and given that the article was published in 1890 clearly does not take into account technical aspects of personal privacy in cellular systems. However, the authors foresaw the need to capture privacy as a generic right and state that the right applies to "any modern device" that can infringe on the personal privacy rights. To round off, the authors recognize that "4. The right to privacy ceases upon the publication of the facts by the individual, or with his consent." Thus, in this framework one can easily capture *location based services*, which can be considered perfectly acceptable provided one has obtained user consent.

The need for a practical privacy mechanism in an all-digital world was also recognized fairly early on. The influential 1981 paper "Untraceable Electronic Mail, Return Addresses, and Digital Pseudonyms" by Chaum [10] is often cited as the starting point for privacy enhancing technologies (PETs). The need for privacy in cellular systems is recognized in the literature [11–14]. Furthermore, it was also recognized that the mechanisms provided in the 2G (and 3G) systems was inadequate. It was also acknowledged that one could benefit from domain separation of knowledge to enhance subscriber privacy [15]. Some of these ideas have been incorporated in work presented in this book while others served as inspiration to improve and enhance the security and privacy of the cellular subscriber.

1.2 Organization and Structure

The book can roughly be divided into two main parts, where Chapters 2 and 3 treat access security in the existing 2G, 3G and soon-to-come 4G cellular systems from the 3GPP. This provides a solid background for understanding the issues, the challenges and the strengths and weaknesses of access security in the current systems. The second part of the book is more explorative. Chapters 4 and 5 provide in-depth discussions of system models and personal privacy aspects, while Chapter 6 provides a set of design guidelines for access security in a cellular system context. Chapter 7 discusses entity authentication and provides an example of a privacy enhanced authentication and key agreement protocol. Chapter 8 goes into the thorny territory of formal verification, which can be quite difficult, but it is an essential part of security protocol design, although the "proof" part of formal verification should not be overplayed.

Overview of the Chapters

1. *Introduction*
   This chapter is the introduction chapter.
2. *Access Security in 2G and 3G Systems*
   In this chapter the GSM/GPRS and the UMTS system architecture is described together with the corresponding access security. Access security in GSM shows clear signs of old and new and it was always limited in scope. Access security for GPRS builds directly on GSM access security and is as such fairly unremarkable. The UMTS access security architecture, with its strong points and weaknesses, is the main basis for the analysis and design of the enhanced access security procedures presented in this book.
3. *Long-Term Evolution*
   The term Long-Term Evolution (LTE) is used by the 3GPP as an official term for its entry into 4G territory. We shall mainly investigate the LTE security architecture in this chapter. Strictly speaking, the currently LTE (3GPP Release 8) system is not a full 4G system according to the definitions from ITU-R (IMT Advanced). The LTE shortcomings, mainly radio related, are already identified and LTE Advanced is planned to be a full 4G solution. The system and security architecture in LTE Release 8 will be retained in LTE Advanced.
4. *Access Security for Future Mobile Systems*
   This chapter examines and describes various aspects of a possible future

mobile access security architecture. A main objective is to investigate assumptions and requirements for a Privacy Enhanced 3-way Authentication and Key Agreement (PE3WAKA) protocol. The discussion takes into account aspects of the target system architecture, cellular environment design constraints and system performance requirements.

5. *Privacy Matters*

The current 3G cellular systems have, despite some shortcomings, achieved a reasonable level of subscriber data confidentiality and data integrity protection over the wireless link. However, when it comes to provision of identity privacy and location privacy the 3G systems have not fared well. The upcoming 4G (LTE) builds on the UMTS subscribers' identity structures and thus the personal privacy problems are still there. Provision of a credible subscriber privacy scheme is therefore thoroughly discussed and proposals for a solution are made.

6. *Principles for Cellular Access Security*

This chapter consists mainly of input from Chapters 4 and 5 and is intended to sum up the requirements and principles derived in those chapters. Instead of presenting a set of concrete design requirements for the access security architecture the requirements have been formulated as a set of design principles.

7. *Authentication and Key Agreement*

The purpose of this chapter is to present the *Privacy Enhanced 3-Way Authentication and Key Agreement (PE3WAKA)* protocols. The subject of authentication and key agreement and what exactly the authentication goals should be is discussed in this chapter. The chapter also includes one example protocol of the PE3WAKA protocol family.

8. *Security Protocol Verification*

In this chapter the field of formal verification of security protocols is investigated. Several methods and tools are presented and a particular tool set (AVISPA) is chosen as the basis for formal verification of the PE3WAKA example protocols.

9. *Summary*

This chapter briefly summarizes the book.

2

Access Security in 2G and 3G Systems

> Better be despised for too anxious apprehensions,
> than ruined by too confident security.
> – *Edmund Burke (1729–1797)*

2.1 Introduction

This chapter describes and examines the access security architecture of 2G and 3G public cellular systems. The chosen example systems are the 2G GSM system and the 3G UMTS system. The GSM and UMTS systems are both specified by the 3GPP and the systems share a number of traits and a common base.

The UMTS 3G system contains a large amount of abbreviations, terms and symbols. The reader is encouraged to consult the official 3GPP vocabulary and abbreviation document, the 3GPP TR 21.905 [1] which is available online.[1] Note that terms that only affect one specification (or a limited set of specifications) are often only defined in the affected specification(s).

2.2 Addresses and Identities

2.2.1 Subscriber Addresses and Identities

The permanent subscriber identity in GSM and UMTS is the International Mobile Subscriber Identity (IMSI). The permanent identity is assigned by the home operator and is specified in TS 03.03 [16] for GSM and for UMTS it is

---

[1] Available at http://www.3gpp.org/ftp/Specs/archive/21_series/21.905/. Use the latest version.

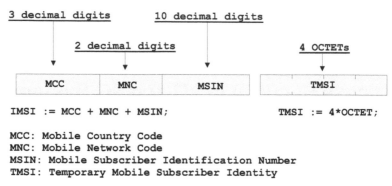

IMSI := MCC + MNC + MSIN;   TMSI := 4*OCTET;

MCC: Mobile Country Code
MNC: Mobile Network Code
MSIN: Mobile Subscriber Identification Number
TMSI: Temporary Mobile Subscriber Identity

Figure 2.1 The IMSI and TMSI identities.

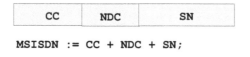

MSISDN := CC + NDC + SN;

CC: Country Code
NDC: National Destination Code (trunk code)
SN: Subscriber Number

Figure 2.2 The ITU-T E.164 compliant MSISDN number.

TS 23.003 [17]. ITU-T Rec. E.212 also describes the IMSI number/identity [18].[2]

The subscriber may also be identified by a temporary identity called the Temporary Mobile Subscriber Identity (TMSI). The TMSI is assigned by the serving network (VLR/MSC or SGSN). The structure of TMSI is specified in TS 03.03 [16] and TS 23.003 [17]. Figure 2.1 depicts the structure of the IMSI and the TMSI. The IMSI and the TMSI are system internal identifiers and should not be confused with the subscriber telephone number. This number, called the MSISDN number (Figure 2.2), is a normal ITU-T Rec. E.164 [19] telephone number. The MSISDN number is a signalling system no. 7 (SS7) address and is used to route calls for ISDN/POTS subscribers.

---

[2] E.212 refers to IMSI as the "International Mobile Station Identity" while GSM TS 03.03 refers to "International Mobile Subscriber Identity". The discrepancy is inconsequential.

2.3 GSM Access Security

2.3.1 In Brief

GSM access security is developed around the need to protect the physical radio access against eavesdropping. The design is based on the GSM Authentication and Key Agreement protocol (GSM AKA) and the use of a symmetric-key stream cipher to provide confidentiality protection between the Mobile Station (MS) and the Base Tranceiver Station (BTS).

All GSM subscribers have a unique identity, the IMSI, and a secret authentication key K_i. Physically, the IMSI and the K_i are located on the SIM card (the smart card) and at the Authentication Centre (AuC). The AuC is co-located with the HLR in the home environment (home network). The IMSI number is not secret *per se*; the Serving Network (SN) will be allowed to learn the IMSI. Furthermore, the IMSI is occasionally transmitted in plaintext by both the SN and the MS over the radio interface. Still, for subscriber privacy reasons, the IMSI exposure should be contained. The authentication key K_i is secret and *only* the SIM and the AuC should ever know the K_i value. The (IMSI, K_i) association is permanent with respect to the subscription.

The GSM AKA is based on a challenge-response mechanism in which the network challenges the SIM with a random challenge (only the network can initiate the GSM AKA sequence). The SIM responds with a signed response. The Message Authentication Code (MAC) based signature is computed over the challenge under control of the secret authentication key K_i. The authentication is not mutual; the network can verify the subscriber identity but the subscriber cannot know if it is in contact with a legitimate network.

During the GSM AKA procedure the 64-bit symmetric secret session key K_c is generated. This key is used by the A5 stream cipher to provide data confidentiality over the access link between the MS and the BTS.

2.3.2 The Scope of GSM Access Security

The primary documents for GSM access security are TS 02.09 "Security aspects" [20] and TS 03.20 "Security related network functions" [5]. The main security requirements can be summarized as follows:

- Provision of subscriber identity confidentiality.
- Provision of subscriber identity authentication.
- Protection of control signalling information elements, confidentiality protection of connectionless user data and provision of data confidentiality for physical connections.

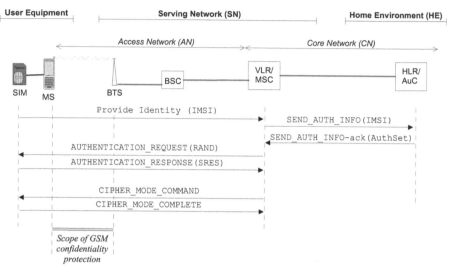

Figure 2.3 Outline of GSM access security.

- It shall[3] be possible to introduce new authentication and ciphering algorithms.
- The security procedures include mechanisms to enable recovery in event of signalling failures.

Figure 2.3 depicts the two-stage GSM AKA protocol. The challenge-response part of the GSM AKA is between the VLR/MSC and the SIM card, and as can be seen the confidentiality protection is between the BTS and the ME. The cipher key K_c must therefore be transferred from the SIM to the ME and from the VLR/MSC to the BTS.

2.3.3 The Cryptographic Functions

The Authentication Set

In GSM one has defined a compound information element called the *Authentication Set*. The Authentication Set, also known as the *triplet*, is produced by the AuC and forwarded to the VLR/SGSN.[4] The Authentication Set consists of the following:

---

[3] The word "shall" has a special meaning in ETSI/3GPP. Use of "shall" indicates a mandatory feature.

[4] VLR/SGSN is a shorthand notation for VLR/MSC and/or SGSN.

```
Authentication Set = {
    RAND : 128 bit; - The (pseudo)random challenge
    SRES :  32 bit; - The signed response
    Kc   :  64 bit; - The session cipher key }
```

Generation of the Random Challenge ($RAND$)

The random challenge, despite its name, need not be truly random. In fact, it is almost certainly generated by an algorithm. The challenge, $RAND$, is always given in the context of a (IMSI,K_i)-tuple. For each authentication set a new $RAND$ must be generated by the AuC. Briefly we assume:

- The value of $RAND$ must never be repeated for an individual subscriber.
  There is no state in the GSM AKA computations and both the *challenge* and *response* are sent in plaintext over-the-air. Thus, the relationship between the challenge and the response is completely fixed.
- Unpredictability of $RAND$
  If prediction is possible the intruder will be afforded the advantage of pre-computation. Therefore it should not be possible to predict the value of the Nth instance of $RAND$ based on knowledge of all previous instances ($[0..N - 1]$) of $RAND$.

The $RAND$ is produced by a pseudo-random number generator, $prng()$, (Equation 2.1). The GSM specifications does not formally define a function for the $prng()$ and the only stated requirement on $RAND$ is that it must be unpredictable [5, section 3.2]).

$$prng(internal\_state) \to RAND \qquad (2.1)$$

The $RAND$ is a 128 bit field and it nominally has a state-space of 2^{128}. Provided that the $prng()$ function yields a uniformly distributed and unpredictable output the state-space of $RAND$ itself is large enough as to leave brute-force guessing attacks infeasible.

The requirement for no repetition is quite easy to accommodate. Algorithms exist that guarantee to produce long cycles without repetition. One class of algorithms is the so-called linear congruential pseudo-random number generators. These generators may be well suited for the type of randomness required by Monte Carlo simulations, etc., but they are not good cryptographic pseudo-number generators. In fact, it has been demonstrated that all polynomial congruential generators can be broken [21].

In practice one does not need a mathematical guarantee that the value of $RAND$ is never repeated. It is adequate to guarantee a sufficiently low probability for a $RAND$ repetition (see [22, remark 10.11]). Inevitably the birthday paradox will apply here. The birthday paradox is a valid assumption since any collision of all the draws (i.e. any pair) constitutes a repetition. That is, after approximately 2^{64} draws there will likely ($p > 0.5$) be at least one collision. However, each $RAND$ draw corresponds to one authentication event (ae), and for the given context the number of valid authentication events will never approach the number 2^{64}.

So how many ae occurrences can there reasonably be for a single subscriber? One may conservatively assume that a specific context can live for at most 10 years (your subscription may still continue beyond this). That is, the lifetime of a SIM card is limited to at most 10 years. Authentication in GSM is not a particularly frequent event. To be very conservative let us assume one authentication event occurs every minute for the duration of the context lifetime. This roughly totals $1 \times 60 \times 24 \times 365 \times 10 = 5256000$ events. Thus: $ae < 2^{23}$.

One practical *prng* approach is to use a counter as input and then apply a bijective function to provide a random mapping to an output space. By definition a cipher algorithm is a bijective function and the standard assumption is that ciphers shall appear to have a random mapping between the output and the input (definitions 7.1 and 7.2 in [22]). Designs like the Fortuna proposal [23], which is based on a block cipher (AES is suggested) in counter-mode (CTR) to produce the pseudo-random stream, should therefore adequately cover our needs.

The Authentication Algorithm (A3)
The external interface to the A3 algorithm is given in annex C.2 in TS 03.20 [5]. Algorithm A3 is used for authentication of a mobile subscriber identity. The A3 function computes the expected response $SRES$ from a random challenge $RAND$ sent by the network. The A3 algorithm is physically located in the SIM and at the AuC. The function is defined in (2.2). The A3 input/output is defined in Table 2.1.

$$A3_{K_i}(RAND) \rightarrow SRES \tag{2.2}$$

TS 03.20 [5] does not specify cryptographic requirements to the A3 algorithm, but the A3 algorithm should obviously be designed to make it computationally infeasible to derive K_i from knowledge of $RAND$ or $SRES$. This property should hold even if the intruder has collected a large number

Table 2.1 Input/output to the A3 function.

Information Element	Length	Secret/Public
K_i	128 bit	Secret (permanent)
$RAND$	128 bit	Public
$SRES$	32 bit	Public

Table 2.2 Input/output to the A8 function.

Information Element	Length	Secret/Public
K_i	128 bit	Secret (permanent)
$RAND$	128 bit	Public
K_C	64 bit	Secret (session)

of $(RAND, SRES)$ pairs. One must also require that it is computationally infeasible to derive $SRES$ based only on knowledge of $RAND$. There is no way for the A3 function to verify the validity of the challenge. The requirements must therefore hold even for a large number of *chosen* $(RAND)$ challenges. In practice the A3 algorithm can be implemented by a MAC function.

The Key Derivation Algorithm (A8)

The external interface to the A8 algorithm is given in annex C.3 in TS 03.20 [5]. The A8 function takes exactly the same inputs as the A3 algorithm. The output from the A8 function is the secret session key K_C. The A8 is physically located on the SIM and at the AuC. The function is defined in (2.3). The input/output is defined in Table 2.2.

$$A8_{K_i}(RAND) \rightarrow K_C \qquad (2.3)$$

The symmetric key K_C is a secret key, but it must necessarily be known to the SN and the ME. TS 03.20 [5] does not specify cryptographic requirements to the A8 algorithm, but the A8 algorithm should clearly be designed such that it is computationally infeasible to derive K_i from knowledge of $RAND$ and K_C. This property should hold even if the intruder has collected a large number of $(RAND, SRES)$ pairs. Obviously, one must also require that it is computationally infeasible to derive K_C based on knowledge of $RAND$ alone.

Table 2.3 Input/output to the A38 function.

Information Element	Length	Secret/Public
K_i	128 bits	Secret (permanent)
$RAND$	128 bits	Public
$K_C \| SRES$	96 bits	Secret‖Public

The Combined GSM AKA Algorithm (A38)

The A3 and A8 algorithms are in reality one function, denoted as the A38 algorithm (Equation 2.4). The input/output is defined in Table 2.2.

$$A38_{K_i}(RAND) \rightarrow K_C \| SRES \tag{2.4}$$

The requirements to A3 and A8 must be strengthened for the combined A38 function. It must be computationally infeasible to derive K_i from knowledge of $RAND$, $SRES$ and K_C. This property should hold even if the intruder has collected a large number of $(RAND, SRES, K_C)$-tuples. Obviously, one must also require that it is computationally infeasible to derive K_C based on knowledge of $RAND$ and $SRES$. One must also retain the requirement for the A3 algorithm that it must be infeasible to derive $SRES$ from knowledge of $RAND$ alone.

As with A3 and A8, there is no way for the A38 function to verify the validity of the challenge. Consequently, the requirements must hold for a large number of chosen challenges. The meaning of large is context dependent; for an intruder with access to the SIM card the requirement should ideally hold for a large proportion of the cycle length of the $RAND$ parameter (which nominally is 2^{128}). For an adversary operating on the air-interface one must take into account that the intruder will only get access to the $(RAND, SRES)$-pair (possibly with chosen $RAND$) and that an over-the-air attack is *much* more time consuming per challenge. The A38 algorithm can be implemented by a MAC function.

The COMP128 Implementations

The A3 and A8 algorithms are considered to be operator specific. In principle there is therefore no need for a standardized solution. Nevertheless the GSM Association has made available reference implementations of the A3/A8 algorithms. The "standard" implementations are collectively known as the COMP128 algorithms. The first of the COMP128 algorithms, now known as COMP128-1, is inherently insecure and breakable in real-time. All COMP128 algorithms implement both A3 and A8. See the short paper on cloning [24] for an overview.

- **COMP128-1:** The "original" COMP128 algorithm. Designed to produce K_C with 54 significant bits. The algorithm is *completely broken* [25] and one is *strongly* advised against using it.
- **COMP128-2:** No serious flaw has been reported for the COMP128-2 algorithm. The algorithm produces a K_C with 54 significant bits.
- **COMP128-3:** This algorithm is identical to COMP128-2, but key K_C now has 64 significant bits.
- **COMP128-4:** Also known as "GSM Milenage". Specified in TS 55.205 [26]. Based on the 3GPP MILENAGE algorithm set used for UMTS. No weaknesses are yet reported.

The A5 Stream Cipher

The external interface to the A5 algorithm is given in annex C.1 in TS 03.20 [5]. The A5 algorithms must be common to all GSM networks and all mobile stations. Currently, there are three fully defined implementations of the A5 algorithm. They are respectively known as the A5/1, the A5/2 and the A5/3 algorithm. A definition is given in Equation (2.5).

$$A5_{K_C}(COUNT) \rightarrow BLOCK1\|BLOCK2 \qquad (2.5)$$

The A5 input/output data:

```
A5 input = {
    Kc    :  64 bit; -- Secret key
    COUNT :  22 bit; -- TDMA frame number (explicit synch.)}

A5 output = {
    BLOCK1 : 114 bit; -- pseudo-random data; (BTS -> MS)
    BLOCK2 : 114 bit; -- pseudo-random data; (MS -> BTS)}
```

The BLOCK data is XORed with the uplink and downlink Normal Burst frames as identified by COUNT. The A5 algorithm must be able to output a set of $BLOCK$ data pr burst; i.e. it must complete one cycle in less than 4.615 ms. The A5 algorithms:

- **A5/1** – This is the standard A5 implementation.
  The algorithm is now considered to be cryptographically broken. The first practical attack was presented by Biryukov, Shamir and Wagner in "Real Time Cryptanalysis of A5/1 on a PC" [27]. The attack cleverly leverages the potential for space/time trade-offs to arrive at a practical attack. The A5/1 algorithm is still widely used.

- **A5/2** – This is the reduced strength version of A5.
  It was designed to be sufficiently weak to allow export to the then
  Eastern Block countries. It is considered to be completely broken, and
  practical attacks have been devised [28]. A5/2 should *not* be used.[5]
- **A5/3** – This version of A5 is derived from the 3G KASUMI
  crypto-primitive. No known practical attack exists against the A5/3
  algorithm.

Secret Algorithms

The GSM algorithms are not publicly available. The same is true for the
GPRS algorithms. Originally, the secrecy was due to export control and the
fact that cryptographic algorithms were officially considered to be "muni-
tion". Export restrictions at the time were an important factor, but today this
is not an issue. Export regimes are regulated by the Wassenaar Arrangement
(http://www.wassenaar.org/). The 3GPP, which now supervises the development of all GSM, GPRS and UMTS security standards and algorithms, does
publish all new 128 bit designs.

2.3.4 Devices and Nodes

Figure 2.3 gives an overview over the nodes and devices involved in GSM
access security.

Subscriber Identity Module (SIM)

The original GSM SIM card is defined in TS 02.17 [29] and TS 11.11 [30].
Table 2.4 gives an overview of the data present on the SIM. The algorithm
A38 is also present on the SIM card. The SIM device, which is a tamper
resistant smart card in compliance with the ISO/IEC 7810 standard [31], contains the permanent subscriber identity (IMSI) and all the security credentials
and algorithms needed for entity authentication.

Mobile Equipment (ME)

The A5 cipher algorithm is located on the ME and in the BTS. The ME must
therefore know the cipher key K_C. Table 2.5 gives an overview of data that
may be present on the ME. The $CKSN$ is an index to the GSM security context. It is provided by the network in the DTAP_AUTHENTICATION_REQUEST
command together with the $RAND$. It also serves to identify the associated
K_C.

---

[5] The official GSM Association network cut-off date was 2006.06.01.

Table 2.4 Security credentials on the SIM.

Data	Size	Type	Comment
$IMSI$	3+2+10 dec. digits	Permanent	The Subscription Identity
K_I	128 bits	Permanent	Pre-shared secret
K_C	64 bits	Temporary, Output	The confidentiality key
$RAND$	128 bits	Temporary, Input	Random challenge
$SRES$	32 bits	Temporary, Output	Signed Response
$CKSN$	3 bits	Temporary	Cipher-key Seq.num.

Table 2.5 Security credentials on the ME.

Data	Size	Type	Comment
$IMEI$	15 dec. digits	Permanent	ME Identity
$TMSI$	32 bits	Temporary	The Local/Temporary Identity
K_C	64 bits	Temporary, Output	The confidentiality key
$CKSN$	3 bits	Temporary	Cipher-key Seq.num.
$IMSI$	3+2+10 dec. digits	Permanent	The Subscription Identity
$RAND$	128 bits	Temporary, Input	Random challenge
$SRES$	32 bits	Temporary, Output	Signed Response

The GSM AKA protocol is not executed for every session/connection setup. During session/connection setup the ME will indicate the current $CKSN$ and if the value matches the $CKSN$ value stored in the network it may decide to postpone execution of the GSM AKA protocol. The network then starts ciphering with the old key indicated by $CKSN$. Section 4.3.2 in TS 04.08 [32] provides details on the handling of $CKSN$.

Base Station System (BSS)

Base-Tranceiver Station (BTS) is the network terminating point of the A5 cipher. The data indicated in Table 2.5 for the ME also apply to the BSS. The Base Station Controller (BSC) does not play a role in GSM security.

Visited Location Register/Mobile Switching Center (VLR/MSC)

The VLR and MSC are defined to be separate nodes, but they are in fact normally implemented as one physical node. Table 2.6 gives an overview of the security related data stored at the VLR/MSC. It is the VLR/MSC that assigns the GSM AKA security context index and it is the VLR/MSC that decides on the number of times the K_C is reused. The MS does have some influence on the number of reuses and it can give a "no key available" indication in $CKSN$. The VLR/MSC must then initiate the GSM AKA protocol to derive a new security context.

Table 2.6 Security credentials known to VLR/MSC.

Data	Size	Type	Comment
$IMEI$	15 dec. digits	Permanent	ME Identity
$TMSI$	32 bits	Temporary	Assigned by VLR/MSC
K_C	64 bits	Temporary, Output	The confidentiality key
$CKSN$	3 bits	Temporary	Assigned by VLR/MSC
$IMSI$	3+2+10 dec. digits	Permanent	The Subscription Identity
$RAND$	128 bits	Temporary, Input	Random challenge
$SRES$	32 bits	Temporary, Output	Signed Response

Table 2.7 Security credentials at the AuC.

Data	Size	Type	Comment
$IMSI$	3+2+10 dec. digits	Permanent	The Subscription Identity
K_I	128 bits	Permanent	Pre-shared secret
K_C	64 bits	Temporary, Output	The confidentiality key
$RAND$	128 bits	Temporary, Input	Random challenge
$SRES$	32 bits	Temporary, Output	Signed Response

Home Location Register/Authentication Center (HLR/AuC)

Formally, the HLR and AuC are separate nodes, but in practice the AuC is integrated with the HLR. Administratively they are however to be treated as separate nodes. The AuC implements the A38 algorithm and a pseudo-random number generator function ($prng()$). It is noted that the $prng()$ function is only present at the AuC. Table 2.7 gives an overview over the AuC security data. The HLR will know all the AuC data except K_i.

Equipment Identity Register (EIR)

The sole purpose of the EIR database is for checking the equipment identity (IMEI). The MSC will check the reported IMEI with the EIR. The check frequency is a matter of operator policy. The operators may log the ($IMSI, IMEI$) association. A Law Enforcement Agency (LEA) can use the log for tracking down stolen handsets and authorize blocking of subscriptions that use stolen handsets.

The IMEI check is by means of the MAP_CHECK_IMEI message. The EIR will respond with one of the following indications:

- White list – The IMEI is verified and the call is accepted.
- Gray list – The call is accepted, but the MSC may activate monitoring of the MS.
- Black list – The call is dropped (should also trigger a IMEI–IMSI logging event).

2.3.5 The GSM AKA Protocol in Alice–Bob Notation

The GSM AKA protocol is outlined in Figure 2.4 in the common Alice-Bob notation. The parties are MS, SN and HE. The sequence starts with a triggering event, which in the example is the initial registration (the location updating signalling messages have otherwise been omitted). The registration event includes identity presentation with IMSI. The message names have been abridged. Observe that there is no standardized protection mechanism for the channel between HE and SN in the GSM environment.

1. $MS \rightarrow SN$: LOC_UPDA_REQ($IMSI$)
2. $SN \rightarrow HE$: SEND_AUTH_INFO($IMSI$)
3. $HE \rightarrow SN$: SEND_AUTH_INFO(($RAND, SRES, Kc$))
4. $SN \rightarrow MS$: AUTH_REQ($RAND, CKSN$)
5. $MS \rightarrow SN$: AUTH_RESP($SRES$)

Figure 2.4 The GSM AKA protocol.

2.3.6 Subscriber Identity Confidentiality

Subscriber identity confidentiality in GSM is based on the $TMSI$ identity. In the basic scheme the subscriber initially is identified by the permanent identity $IMSI$. This identification is in clear. Subsequent to the initial presentation the network will normally run the GSM AKA protocol. During execution of the AKA protocol the $IMSI$ is verified and the security context is established. The SN can then start ciphering and assign the local temporary identity $TMSI$. The $TMSI$ is provided in ciphertext form and will later only be used in plaintext form. The association between the $IMSI$ and the $TMSI$ will thus only be known to the MS and the SN.

However, the scheme has weaknesses. It is inadequate against an active adversary. The network, or an adversary masquerading as the network, may page the subscriber with $IMSI$ in clear and the network may request the MS to provide the $IMSI$. Even without an active adversary the $IMSI$ invariably and regularly will be exposed over the Um-interface (this is the over-the-

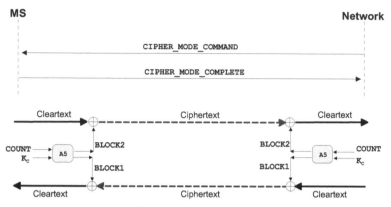

Figure 2.5 The use of A5.

air interface). With respect to anonymous subscriber tracking, it is observed that the $TMSI$ scheme does not prevent tracking unless the $TMSI$ value is frequently re-assigned. This is possible to do, but there is no requirement to do so. The GSM subscriber identity confidentiality scheme is specified in TS 03.20 [5, section 2]. It is noted that UMTS has inherited the GSM subscriber identity confidentiality scheme.

2.3.7 Subscriber Data and Subscriber Signalling Data Confidentiality

Scope

The confidentiality protection can only be initiated by the network. The link between the MS and the BTS is protected by the A5 stream cipher to provide data confidentiality of all user data. The scope of protection includes all user specific signalling and all user data transmitted over the Um-interface. Figure 2.5 depicts the use of the A5 algorithm. As is shown, the uplink and downlink streams are protected with separate pseudo-random key streams ($BLOCK1, BLOCK2$). The $BLOCK$s are each 114 bit wide and matches the *Normal Burst* which also has 114 bits of payload.

The encryption is performed by bitwise modulo two addition (exclusive or) of the plaintext and the key stream. Decryption takes place by executing the bitwise modulo two addition on the ciphertext and the key stream. The use of the TDMA frame number ($COUNT$) ensures that synchronization is maintained.

Synchronization and the TDMA Frame Number

The encryption/decryption is synchronized with the TDMA frame numbers. As a consequence the period of the cipher is identical to the period of the TDMA hyperframes. Unfortunately this cycle length is shorter than desirable. The hyperframe will last for $26 \times 51 \times 2048$ bursts. In GSM this means that the key stream will be exhausted after approximately 3 hours and 29 minutes. The problem with this is that the security is drastically weakened when two different user data frames are encrypted with the same pseudo-random key stream. This highlights three problems with GSM security:

- There is no automatic re-keying during connections in GSM.
- Connections are allowed to outlive the cycle length of the A5 algorithm.
- The key K_C is allowed to be reused for several connections.

Cipher Processing

Another problem with GSM security is that ciphering is applied *after* the code expanding error corrections codes have been added (see Figure 2.6). This design is embarrassing as it has been known for a long time that one should ideally *remove* redundancy from the input. The classical reference here is C.E. Shannon's seminal 1948 paper "Communication Theory of Secrecy Systems" [33].[6]

Algorithm Negotiation and Missing Key Binding

Prior to the ciphering the MS may indicate its capabilities to the network. The mechanism is generic and many different MS capabilities can be signalled to the network. The MS can convey to the network which cipher algorithms it supports. The network will compare the MS capabilities with its own list of supported algorithms and select a preferred one. Currently, the MS may indicate A5/1, A5/2 and A5/3. The network may request either one (or none) of those algorithms. A5/2 is deprecated and should *not* be chosen.

In GSM there is no binding between the key K_C and the selected algorithm. Neither are there any explicit key expiry conditions and the security setup sequence leaves the MS with limited influence on the rekeying policy. The missing key binding is very problematic and has become an Achilles heel of GSM security. In the Barkan, Biham and Keller (BBK) paper, "Instant Ciphertext-only Cryptanalysis of GSM Encrypted Communication" [28], the authors ruthlessly expose GSM security. One of the described attacks cleverly

---

[6] Shannon calls for the use of a *transducer* to remove redundancy several places in the paper.

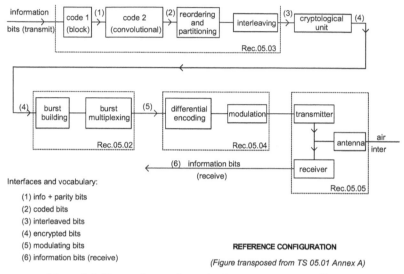

Figure 2.6 Encryption performed subsequent to error coding.

exploits the preposterously weak A5/2 algorithm to derive the key K_C. With no key binding it does not help if the legitimate SN operator subsequently orders the MS to use the safe A5/3 algorithm since the broken key will be used.

2.4 The UMTS System Architecture

2.4.1 The Basic Architecture

The basic UMTS Release 99 access security architecture is defined in Technical Specification (TS) 33.102 [34]. The scope for Release 99 was strictly limited to access security.

The basic UMTS system architecture is quite similar to the GSM/GPRS architecture. The obvious difference between the two architectures is the UMTS Access Network. The Universal Terrestrial Radio Access Network (UTRAN) provides a completely new radio system based on Wideband-CDMA and the lower layers are very different from the GSM TDMA-based radio system. The description of the UMTS system architecture given here is brief. Numerous books on the topic exist, but most books either spread their focus too much or they are too specialized (more often than not on the radio system). Figure 2.7 presents an overview and depicts the main security-relevant parts of the UMTS architecture.

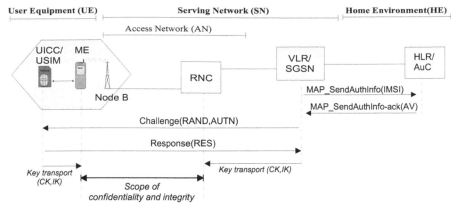

Figure 2.7 The UMTS access security architecture.

2.4.2 The User Equipment

The term MS has been replaced by User Equipment (UE) in UMTS (TR 21.905 [1]):

User Equipment: A device allowing a user access to network services. For the purpose of 3GPP specifications the interface between the UE and the network is the radio interface. A User Equipment can be subdivided into a number of domains, the domains being separated by reference points. Currently defined domains are the USIM and ME Domains.

The USIM domain consists of the USIM application running on the UICC (smart card). According to TR 21.905 [1]:

UICC: a physically secure device, an IC card (or "smart card"), that can be inserted and removed from the terminal equipment. It may contain one or more applications. One of the applications may be a USIM.

Universal Subscriber Identity Module (USIM): An application residing on the UICC used for accessing services provided by mobile networks, which the application is able to register on with the appropriate security.

The UICC is specified in TS 31.101 [35] and the USIM is specified in TS 31.102 [36]. The USIM is functionally analogous to the SIM in GSM, and the USIM is the entity which holds the subscriber identity (IMSI) and the security credentials (including the AKA functions). The Mobile Equipment (ME) contains terminal functionality and the radio termination. The ME–UICC/USIM interface is similar to the ME–SIM interface.

2.4.3 The Access Network (AN)

The UMTS access network (UTRAN) consists of two main nodes; the Radio Network Controller (RNC) and the base station (NodeB). One of the differences between a UMTS RNC and a GSM BSC is that the RNC does play an important role in the access security architecture. The UMTS link layer protection terminates in the RNC, and thus the RNC will contain the $f8$ and $f9$ security functions. The BTS plays an important role in the GSM security architecture, but the NodeB does not have a similar role in the UMTS security architecture. Figure 2.7 illustrates the roles of the nodes with respect to security.

2.4.4 The Core Network (CN)

The UMTS CN is an evolved version of the GSM/GPRS CN. A UMTS CN is also able to support GSM/GPRS access networks which, when connected to UMTS CN, are called GERAN. From an access security point of view there are relatively few changes between a GSM/GPRS CN and a UMTS CN, but we note that while the 2G GPRS SGSN was the termination point for link layer encryption (the GEA cipher function is located in the SGSN node in GPRS) the termination point for packet services in UMTS is in the RNC. It is also noted that in UMTS (Release 5 and onwards) one tends to refer to the Home Subscriber Server (HSS) instead of HLR/AuC. The HSS can be thought of as a logical unit consisting of HLR, AuC and possibly an AAA[7] server. For the sake of convenience one often refers to "VLR/SGSN" when ones mean to indicate both VLR/MSC and SGSN.

2.4.5 System Addresses and Identities in UMTS

Initially, new identities and address structures were defined for UMTS. However, since UMTS Release 99 relies on the GSM signalling protocols and still uses the SS7 signalling system it became apparent that there was little to be gained by changing the identities and addresses. In fact, the opposite would be

---

[7] Authentication, Authorization and Accounting (AAA) defines an internet architecture framework and protocol suite for authentication, authorization and accounting purposes.

true since one could then potentially run into compatibility problems. Compatibility with GSM was a major design goal and so it was decided to keep the GSM/GPRS identities and addresses. The identity structures (IMSI, MS-ISDN, IMEI, etc.) in UMTS are therefore exactly the same as in GSM/GPRS. TS 23.003 [17] is the definitive reference on UMTS numbering, addressing and identification schemes.

2.5 Scope and Requirements for UMTS Access Security

2.5.1 Scope of the UMTS Access Security Architecture

The access security mechanisms found in GSM was the starting point for UMTS access security. Of course, the design objectives for the UMTS security architecture was not limited to the existing solutions in GSM. The high level goals and principles of UMTS access security are captured in TS 33.120 [37]. The main principles are:

- *UMTS security will build on the security of second generation (2G) systems*
  That is, existing GSM security features that are needed and robust shall be kept.
- *UMTS security will improve on the security of 2G systems*
  UMTS security will address and correct real and perceived weaknesses in second generation systems. This includes the introduction of mutual authentication and strong encryption with 128-bit key length.
- *UMTS security will offer new security features*

2.5.2 The Security Requirements

The UMTS security requirements are found in chapter 8 in TS 21.133 [38]. The detailed requirements are formulated to be in line with the security objectives and principles [37] and to meet the high-level security requirements. The security requirements are also devised to address the major and medium level threats identified in chapter 7 of TS 21.133.

2.6 UMTS Access Security Functions and Procedures

The purpose of the UMTS AKA protocol is to establish an authenticated security context between the USIM and the VLR/SGSN. The UMTS AKA protocol is divided into two phases; one phase where security credentials

```
IMSI :      8 byte; -- Permanent subscriber identity,
                     16 (hexa)decimal digits
K    :    128 bit;  -- Long-term pre-shared secret
```

Figure 2.8 Permanent UE-HE security related data.

are forwarded from HLR/AuC in the home network to the VLR/SGSN in the serving network and one phase which consists of a challenge-response exchange between the USIM and the VLR/SGSN. The challenge-response protocol is a single-pass protocol which provides mutual entity authentication and key agreement. The UMTS security architecture, and the UMTS AKA protocol in particular, are presented in several reports, papers and books, see e.g. [39–41].

2.6.1 Security Credentials and Parameters

The cryptographic basis for the authentication is a 128 bit pre-shared secret key, K, which resides exclusively in the USIM and the AuC. The permanent subscriber identity $IMSI$ and K are the security credential basis for the UMTS AKA protocol. Figure 2.8 depicts this.

Each UMTS AKA protocol invocation is associated with a session-oriented security credential set called an Authentication Vector (AV). The AV is defined in Definitions 2.6 and 2.7 and is depicted in Figure 2.9. The full session-oriented security context also contains the $IMSI$ and (when applicable) the temporary local identity $TMSI$ (see the beginning of this chapter for more on $IMSI$ and $TMSI$).

$$AV := (RAND, XRES, CK, IK, AUTN) \qquad (2.6)$$

$$AUTN := (SQN \oplus AK)\|AMF\|MAC{-}A \qquad (2.7)$$

The AV corresponds to the GSM triplet, but contains more elements to facilitate mutual authentication. The AVs contain sensitive data and it is therefore clear that the transfer of AVs between the HLR/AuC and the SGSN/VLR needs to be secured against eavesdropping and modification. Observe that the use of the sequence number in the $AUTN$ may allow an intruder to track the subscriber. To preserve anonymity, the SQN value may therefore be concealed by XORing the value with an anonymity key AK. The AK is generated by the $f5$ function.

```
Authentication Vector = AV = {
    RAND  :  128 bit; -- The random challenge
    XRES  :  32-128 bit; -- The expected response
    CK    :  128 bit; -- The confidentiality key
    IK    :  128 bit; -- The integrity key
    AUTN  :  128 bit; -- The authentication token}

Authentication Token = AUTN = {
    SQN   :   48 bit; -- The sequence number
    AMF   :   16 bit; -- The authentication management field
    MAC-A :   64 bit; -- Signature to authenticate the challenge}
```

Figure 2.9 The authentication vector (AV).

2.6.2 The Cryptographic Functions in UMTS

Overview

The UMTS AKA functions and parameters are defined in TS 33.105 "Cryptographic algorithm requirements" [42]. The UMTS AKA functions includes a pseudo-random number function to construct the $RAND$ parameter, a MAC function to sign the challenge, a function to compute the response and a set of key derivation functions. Additionally, there is a set of functions associated with the sequence number scheme.[8]

The pseudo-random number generating function ($f0$) is only present in the AuC. The cryptographic functions ($f1–f5*$) used in the AKA procedure are located in the USIM and the AuC. The UMTS operators are free to choose any algorithm they want provided it complies with the function input/output specification given in TS 33.105 [42]. The 3GPP has nevertheless developed an example set of functions called the MILENAGE algorithm set [43–47]. The formal status of MILENAGE is that it is provided as an example algorithm set, but in practice it is the default algorithm set for the AKA functions. MILENAGE itself is built around the Rijndael block cipher (Rijndael was later adopted as the AES [48]).

TS 33.105 also specifies the confidentiality function ($f8$) and the integrity ($f9$) function. The $f8$ and $f9$ functions must be fully standardized since the functions are placed in the ME and in the RNC. One anticipated the need for multiple alternative algorithms, and this is captured with the UEA and UIA algorithm identifiers. The UEA and UIA identifiers allow for encoding of

---

[8] The sequence management scheme is not fully standardized. TS 33.102, annex C [34], contains an outline of example schemes.

Table 2.8 UMTS cryptographic functions.

Function	Purpose/Usage	Status	Location
$f0$	Random challenge generating function	OS	AuC
$f1$	Network authentication function	OS (M)	USIM, AuC
$f1*$	Re-synchronization message authentication function	OS (M)	USIM, AuC
$f2$	User authentication function	OS (M)	USIM, AuC
$f3$	Cipher key derivation function	OS (M)	USIM, AuC
$f4$	Integrity key derivation function	OS (M)	USIM, AuC
$f5$	Anonymity key derivation function for normal operation	OS (M)	USIM, AuC
$f5*$	Anonymity key derivation function for re-synchronization	OS (M)	USIM, AuC
$f8$	UMTS encryption algorithm	FS (K,S)	ME, RNC
$f9$	UMTS integrity algorithm	FS (K,S)	ME, RNC

OS – Operator Specific; FS – Fully Standardized
M – MILENAGE; K – KASUMI; S – SNOW-3G

up to 16 different algorithms for both data confidentiality and data integrity. This includes the provision for NULL functions, although the use of a NULL functions for $f9$ (integrity) is not permitted. Up to now two algorithm sets have been developed. The first one is based on the KASUMI block cipher and the second set is based on the SNOW-3G stream cipher.

A set of AKA conversion functions (for backwards compatibility with GSM) is defined in TS 33.102. The conversion functions are described in Section 2.7.2. Table 2.8 depicts the UMTS cryptographic functions and their use.

The Random Challenge Generating Function ($f0$)

In UMTS the *challenge*, which includes $RAND$, is explicitly authenticated (by $MAC\text{-}A$). Furthermore, the SQN parameter provides protection against *challenge* replay attacks. Nevertheless, one ends up with requirements similar to those for GSM (Section 2.3.3). The $f0$ function is only present at the AuC. The MILENAGE algorithm set does not contain the $f0$ function. The 3GPP2/CDMA2000 system, which uses a slightly modified UMTS AKA protocol, does specify a pseudo-random number generating function [49].

$$f0(internal\_state) \rightarrow RAND \qquad (2.8)$$

The Network Authentication Function ($f1$)

The network authentication function (2.9) should be a MAC function [42]. It is required that it be computationally infeasible to derive K from knowledge of $RAND, SQN, AMF$ and $MAC\text{-}A$. Obviously, it must also be infeasible to derive $MAC\text{-}A$ from knowledge of $RAND, SQN$ and AMF alone.

$$f1_K(SQN, RAND, AMF) \rightarrow MAC-A \qquad (2.9)$$

The $f1$ function is used to verify the authenticity of the challenge ($RAND, AUTN$). The USIM, upon receiving the challenge, can verify that the challenge data originated with an entity which possesses K. Given that the USIM and the AuC are the only entities that know K, the USIM is assured that the challenge is authentic.

The Re-synchronization Message Authentication Function ($f1*$)

The USIM must assure itself that the challenge is valid (as in fresh/recent). This is verified with the aid of the sequence number (SQN) included in the challenge. The USIM maintains state information of the present and past SQN values. The exact procedure of SQN verification has not been standardized, but it may be based on a sequence number scheme with some suitable window size for the permissible SQN values. If the challenge is found to be expired the USIM must initiate the re-synchronization procedure.

The re-synchronization message authentication function (2.10) should be a MAC function [42]. It is required that it be computationally infeasible to derive K from knowledge of $RAND, SQN_{MS}, AMF*$ and $MAC\text{-}S$. Similar to the $f1$ function, it must also be infeasible to derive $MAC\text{-}S$ from knowledge of $RAND, SQN$ and AMF alone.

$$f1*_K (SQN_{MS}, RAND, AMF) \rightarrow MAC-S \qquad (2.10)$$

$$AUTS := (SQN_{MS} \oplus AK)\|MAC-S \qquad (2.11)$$

The $f1*$ function is used when the USIM has verified the challenge ($RAND, AUTN$) as being authentic, but also found that it failed SQN verification. The USIM then computes $MAC\text{-}S$ over the SQN_{MS}, the $RAND$ and $AMF*$. The $AMF*$ is simply a default value for AMF (all zeroes). The computation of $MAC\text{-}S$ and the construction of $AUTS$ (Definition 2.11) is described in TS 33.102 [34, chapter 6.3.3]. The re-synchronization procedure itself is also specified in TS 33.102 [34, chapter 6.3.5].

When the HLR/AuC receives the $AUTS$ it checks to see whether $AUTS.SQN_{MS}$ is within the range of the current SQN_{HE}. If so, the

HLR/AuC forwards a set of AVs to the VLR/SGSN based on the current SQN_{HE} settings. Otherwise, the HLR/AuC verifies that the $AUTS$ is authentic and then aligns the SQN_{HE} with SQN_{MS}. It is noted that AK produced for use with $AUTS$ is computed with the $f5*$ function. As was the case for $AUTN$, the use of AK to conceal SQN in $AUTS$ is optional. However, the AK and the $f5*$ function is implemented in the MILENAGE algorithm set [43]. Finally, it is noted that $f1$ and $f1*$ must be independent functions and that use of $f1*$ shall not be permitted to compromise the use of the other AKA functions.

User Authentication Function ($f2$)
The user authentication function (2.12) should be a MAC function [42]. It is required that it be computationally infeasible to derive K from knowledge of $RAND$ and RES. It must also be infeasible to derive RES from knowledge of $RAND$ alone. The requirement must hold even when a large set of $(RAND, RES)$ pairs have been observed by an intruder.

$$f2_K(RAND) \rightarrow RES \qquad (2.12)$$

Key Derivation Functions
There are two main key derivation functions; the cipher key derivation function $f3$ and integrity key derivation function $f4$. It is required that it be computationally infeasible to derive K from knowledge of $RAND$ and CK and/or IK. It must also be infeasible to derive CK and/or IK from knowledge of $RAND$ alone. The requirement holds even if a large set of $(RAND, CK, IK)$ tuples have been observed. The CK and IK keys are secret, but it will be known to the ME and the SN (VLR/SGSN and RNC). The USIM computes CK and IK during the authentication sequence when it receives an authentic and fresh challenge.

$$f3_K(RAND) \rightarrow CK \qquad (2.13)$$

$$f4_K(RAND) \rightarrow IK \qquad (2.14)$$

Anonymity Key Derivation Function ($f5$) (for normal operation)
The anonymity key is only 48 bit long and is used to conceal the SQN value. It is required that it be computationally infeasible to derive K from knowledge of $RAND$ and AK. It must also be infeasible to derive AK from knowledge of $RAND$ alone. The AK key, when generated by the $f5$ function, is only used during construction of the AV, and it shall not be known to

any entity but USIM and AuC. Note that the USIM must derive AK *before* it attempts to verify the $AUTN.SQN$ value.

$$f5_K(RAND) \rightarrow AK \tag{2.15}$$

The specifications (TS 33.102 [34] and TS 33.105 [42]) are somewhat unclear on the use of AK and $f5$ (mandatory or not mandatory to use $AK/f5$). However, TS 33.102 [34], at the end of section 6.3.2, does state that the $f5$ function is allowed to be a NULL function and that AK correspondingly shall be set to zero when SQN concealment is not needed. The MILENAGE algorithm set [43] implements a full $f5$ function.

Anonymity Key Derivation Function ($f5*$) (Re-synchronization)

It is required that it be computationally infeasible to derive K from knowledge of $RAND$ and AK. It must also be infeasible to derive AK from knowledge of $RAND$ alone. The AK key, when generated by the $f5*$ function, is only to be used during re-synchronization. In TS 33.102 [34], at the end of section 6.3.2, it is stated that the normal $f5$ function is allowed to be a NULL function, but same is not said for the $f5*$ function. The MILENAGE algorithm set [43] implements a full $f5*$ function. Finally, it is noted that $f5$ and $f5*$ must be independent functions if SQN concealment is to be guaranteed.

$$f5*_K(RAND) \rightarrow AK \tag{2.16}$$

The MILENAGE Algorithm Set

The MILENAGE algorithm set includes an example set of functions for the $f1\ldots f5*$ functions. The MILENAGE algorithms were constructed by the ETSI SAGE group and MILENAGE is fully defined in publicly available specifications [43–46]. A report and summary of MILENAGE is given both in [50] and in [47].

The actual MILENAGE algorithm set was designed around a kernel function consisting of a block cipher primitive with a block size of 128 bit and under control of a 128 bit key. The particular crypto-primitive chosen was the Rijndael cipher, which later became the Advanced Encryption Standard (AES) cipher [48]. The rationale for choosing Rijndael was that it was well understood and scrutinized as it was (at the time) one of the five AES finalist candidates and that it was particularly well suited to be implemented on a smart card. It was also noted in TR 33.909 [50] that it was seen as an advantage that Rijndael was IPR free.

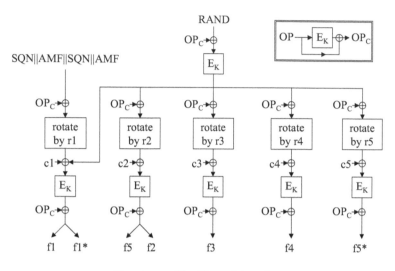

Figure transposed from TS 35.206, Annex A

Figure 2.10 The MILENAGE construction.

As requested by 3GPP the MILENAGE functions (Figure 2.10) implement an operator variant algorithm configuration field (OP). The implementation is interesting and deviates somewhat from the original requirement. The 128 bit wide OP parameter is assumed to be secret, but it must be known to all USIMs within one HE in the basic scheme. The ETSI SAGE task force therefore proposed to use a derived parameter OP_C instead of using the OP field directly. The USIM would then only store the OP_C field. The construction of OP_C is such that even with knowledge of both OP_C and K (including knowledge of the encryption function), the parameter OP is not leaked (Definition 2.17). Thus, the complete compromise of any one USIM would not lead to OP being compromised. This is potentially important since an operator cannot guarantee to protect all USIMs from physical intrusion and compromise, and even a single compromised USIM would have leaked the OP had not the OP_C scheme been used.

$$OP_C := OP \oplus E_K(OP) \tag{2.17}$$

The $f2$ function, which computes the response value (RES), is specified to return a value of between 32 and 128 bit [42]. For MILENAGE the size of RES is fixed to 64 bit [44]. The MILENAGE "mode of operation" construction seems to be well designed and no practical weakness has been reported. A thorough discussion of MILENAGE is given in [41, chapter 8].

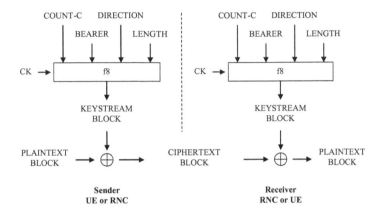

Figure transposed from TS 33.105 (Figure 5)

Figure 2.11 The f8 (confidentiality) function.

The UMTS Confidentiality ($f8$) Function

The mechanism for data confidentiality is defined in [34, 42]. The confidentiality function $f8$ is defined as a stream cipher, which takes as input a bearer identifier ($BEARER$), a direction indication ($DIRECTION$), a time dependent input ($COUNT\text{-}C$), and a length parameter ($LENGTH$). The $f8$ function operates under control of a 128 bit key CK. The $f8$ function is defined in (2.18).

$$f8_{CK}(BEARER, COUNT\text{-}C, DIRECTION, LENGTH)$$
$$\rightarrow KEYSTREAM \tag{2.18}$$

Figure 2.11 illustrates the use of $f8$ to encrypt plaintext by applying a keystream using a bitwise XOR operation. The plaintext is recovered by generating the same keystream using the same input parameters and applying it to the ciphertext using a bitwise XOR operation.

A criticism of the encryption scheme in GSM was that the A5 encryption was applied to the data *after* the data expansion caused by the error correcting/detecting codes. This problem has been rectified in UMTS. As shown in Figure 2.12 no redundancy is added before the encryption.

The UMTS Integrity ($f9$) Function

The mechanism for data integrity of signalling data is defined in TS 33.105 [42] and further described in [34]. The input parameters to the algorithm are

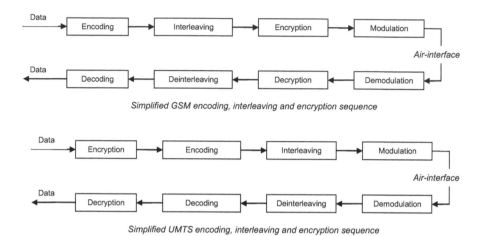

Figure 2.12 Encryption, encoding and interleaving in UMTS.

the integrity key (IK), a time dependent input $(COUNT\text{-}I)$, a random value generated by the network side $(FRESH)$, the direction bit $(DIRECTION)$ and the signalling data $(MESSAGE)$. The output is the message authentication code for data integrity $(MAC\text{-}I)$ which is appended to the message when sent over the radio access link. The receiver computes $XMAC\text{-}I$ on the messages and compares the integrity check values to verify that the message has not been altered. The $f9$ function is defined in Equation (2.19). Figure 2.13 illustrates the use of the $f9$ function.

$$f9_{IK}(COUNT\text{--}I, MESSAGE, DIRECTION, FRESH) \rightarrow MAC\text{--}I$$
$$(2.19)$$

Integrity protection in UMTS is limited to coverage of system signalling messages between the MS and the RNC. The fact that user data is not integrity protected represents a problem under certain circumstances. In particular, there will be situations where encryption is not available and where the integrity protection is the only line of defense.

Birthday paradox attacks must always be considered for MAC/hash functions. However, just finding a collision does not really amount to more than a denial-of-service attack. The trick is to find a meaningful collision. The task of finding meaningful collisions (2nd pre-image) is substantially harder than finding a collision (the Birthday Paradox no longer applies). Furthermore, the $f9$ function is intended to be used exclusively for protection of real-time signalling messages with short expiry periods. It is also unlikely that

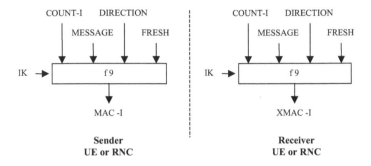

Figure transposed from TS 33.105 (Figure 6)

Figure 2.13 The f9 (integrity) function.

a successful attack can be mounted without falsifying several consecutive signalling messages.

The actual *MAC-I* value used is only 32 bits wide. This would under most circumstances be considered insufficient. In real life, a balance must be struck. For the generally very short signalling messages, even a 32-bit MAC increases the signalling load substantially. Provision of longer MAC values would even make some signalling messages exceed the layer two frame size. The message would then have to be segmented in order to be transmitted. For time critical sequences this is not always permissible. Therefor the choice of only a 32-bit MAC value may be considered sufficient even if the length of the checksum is dangerously small.

The KASUMI Cryptographic Algorithm

The UMTS security architecture allows for 16 different encryption algorithms and 16 different integrity algorithms specified through the UMTS Encryption Algorithm (UEA) identifier and the UMTS Integrity Algorithm (UIA) identifier. The default standard algorithms are identified as UEA1 and UIA1.

The cryptographic core of the UEA1/UIA1 algorithms is based on the KASUMI block cipher [51]. KASUMI is a Feistel cipher with eight rounds and it operates on 64-bit data blocks under control of a 128-bit key. The KASUMI algorithm is derived from Mitsubishi Electric Corporation's MISTY1 algorithm. The design guidelines for MISTY1 were to base it firmly on mathematical/numerical properties, to allow it to be reasonably fast in software on any processor and finally to be fast in hardware. MISTY1 was designed to be provably secure against differential and linear crypto-analysis and the design is built around small components with known resistance against these two types of attacks. These components are then recursively used in the Feistel network. The S-boxes (S7 and S9) are designed to have minimal average differential/linear probability, to be efficient in hardware and to have a high non-linearity order.

To ensure that the quality of the KASUMI cryptographic core and the functions $f8$ and $f9$ were as high as possible, a separate evaluation procedure was carried out. The reports from the external evaluators led to some improvements, but the basic design was not changed. The general conclusion was that the KASUMI algorithms were based on sound design principles and that no practical attacks were found for use within the UMTS context.

UMTS Encryption Algorithm 1

The $f8$ function is defined as a stream cipher. To make the KASUMI primitive fit a modified Output-Feedback Mode (OFM) construction is used. The use of a block cipher in OFM for building a stream cipher is common practice [22, chapter 7.2]). As noted in [39], the OFM construct has some known problems associated with it. In particular, there is a risk that the cycles produced can be too short. It is very undesirable to have the keystream generator enter exactly the same state during a session since that will lead to output of a predictable keystream block.

The $f8$ algorithm takes the $COUNT\text{-}C$, $DIRECTION$ and $BEARER$ identifiers from the radio link as input and the $f8$ output is consequently directly synchronized with the bitstream as it appears to the radio system. Figure 2.14 depicts the UEA1 OFM-based construct.

The Initialization Vector (IV) used in UEA1 is constructed by concatenating the identifiers $COUNT\text{-}C$, $BEARER$ and $DIRECTION$. To create a 64-bit IV, the remaining 26 bits are all set to zero. The IV is used to provide sufficient initial "uniqueness" to the keystream generator. The IV (Definition 2.20) is apparently predictable, which is undesirable. However, the IV is used only once for each cipherstream and the IV is completely independent of the plaintext/ciphertext of any preceding cipherstream. Furthermore, the IV is encrypted by one round of KASUMI using a modified encryption key. The key used for encrypting the IV is CK XORed with a key modifier constant KM. The net effect is that it is very difficult to analyze and predict the pattern of the used IV based on observations of the radio

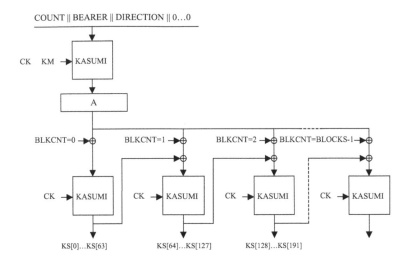

Figure transposed from TS 35.201 (Figure 1, Annex A)

Figure 2.14 The UEA1 construction (keystream generator).

related parameters $COUNT\text{-}C$, $BEARER$ and $DIRECTION$.

$$IV := COUNT\text{-}C \| BEARER \| DIRECTION \| ZERO(26bit) \quad (2.20)$$

UMTS Integrity Algorithm 1

The integrity algorithm $f9$ (UIA1) computes a 32-bit MAC based on the input message under control of an integrity key IK. The UIA1 algorithm is based on a Cipher Block Chaining (CBC) construct. This is a common way of construing MAC functions from block ciphers. Figure 2.15 depicts the UIA1 construction. As one would expect, attacks based on the birthday paradox are possible, but brute-force only attacks is not a real problem. One problem with using a primitive like KASUMI is that the internal state (the block size) is only 64 bits wide. Thus one may then expect ($p > 0.5$) to find a collision after 2^{33} messages. The KASUMI UIA1 cleverly avoids this problem by chaining the XOR of the outputs from all block computations to the input of the final KASUMI computation. A thorough discussion of the KASUMI cipher and the UEA1 and UIA1 constructs is found in [41].

The SNOW-3G Cryptographic Algorithm

The KASUMI cipher was found to be sound and well fitted for its intended purpose, both by the designers and the external evaluators [52]. However,

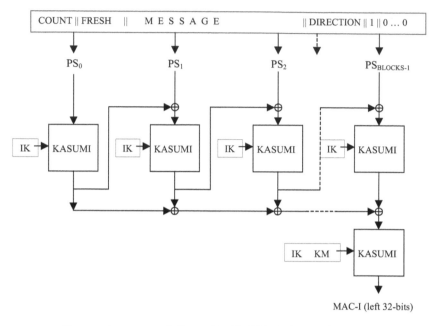

Figure transposed from TS 35.201 (Figure 2, Annex A)

Figure 2.15 The UIA1 construction.

while KASUMI is provably secure against certain attacks there was also a warning that one could not guarantee that it would withstand future attacks. Up to now no practical attack against KASUMI has been published, however, the 3GPP SA3 work group still decided to commission a second cipher suite for the $f8$ and $f9$ functions. To design, adapt and deploy a new standard cipher in the field takes several years. The KASUMI cipher is safe, but should a new attack be found then it would take several years to deploy a replacement algorithm.

The ETSI SAGE group was therefore commissioned to find a suitable existing cipher primitive and adapt it for use with the UMTS $f8$ and $f9$ functions. It was seen as beneficial that the new cipher be based on different cryptographical principles from that of KASUMI. The choice was to base the new algorithm on the streamcipher SNOW-2.0. The new algorithm, called SNOW-3G, was designed with support from the SNOW-2.0 designers. The decision to use SNOW-2.0 as the basis for UEA2 and UIA2 was based on the fact that the algorithm had been available for some years, the algorithm was based on sound cryptographic principles, a lot of public analysis has been undertaken, and no major flaws have been identified. A full description of SNOW-3G and the UEA2 and UIA2 constructs are found in [53–56]. The design and evaluation report is found in [57].

2.6.3 The UMTS Authentication and Key Agreement Protocol

The UMTS AKA protocol is specified in TS 33.102 [34] and discussed in [39, 41].

The Design Context

Originally there were two alternative AKA schemes being developed by the SA3 work group. One is the sequence number based scheme that became the UMTS AKA protocol and the second was a mutual challenge-response scheme.

The mutual AKA scheme was in many ways a better solution and it avoided reliance on a sequence number scheme. However, the mutual AKA scheme would have required message flow changes respective to the GSM AKA protocol. To remain compatible with the GSM AKA message flow was seen as very desirable and eventually the protocol with the sequence number based scheme was chosen as the UMTS AKA protocol.

An added bonus to this approach is that one may run UMTS AKA unmodified over GERAN connections. Thus, provided the subscriber has a UICC/USIM and the VLR/SGSN is 3G compatible, it is possible to have mutual UMTS authentication established over a GSM/GPRS channel.

The UMTS AKA Entities

There are three principal entities in the UMTS AKA sequence. These are:

- **USIM**
  The USIM, which is an application residing on the UICC (smart card), is the user party in the challenge-response sequence. The USIM will forward the session keys (CK, IK) to the ME (mobile phone/device) subsequent to UMTS AKA execution.
- **Serving Network (SN)**
  The SN is the Serving Network and as such a collection of physical nodes and services. The principal SN node in the UMTS AKA execution is the VLR/SGSN server, which is the network party in the challenge-response sequence. The RNC also plays a role in that it is the forwarding point for the session keys (CK, IK).
- **Home Environment (HE)**

The HE is the Home Environment and as such a collection of physical nodes and databases. The principal HE node in the UMTS AKA execution is the HLR/AuC. For UMTS Release 5 and onwards the HE node may be the HSS (Home Subscriber Server). The HSS contains a superset of HLR/AuC functions.

Construction of the Authentication Vector (AV)

The AKA functions have already been described (Section 2.6.2), but not the overall AV construction. Figure 2.16 depicts the construction procedure. The basis for the AV is the random challenge, $RAND$, which is produced at the AuC. Another basic parameter is the $AUTN.SQN$ parameter. The actual construction of the SQN IE will depend on the choice of sequence number management scheme. A proposal for SQN management is given in TS 33.102, annex C [34], but the operators may use other schemes.

1. $f0(\cdot) \rightarrow RAND$
2. $f_{SQN}(\cdot) \rightarrow SQN$
3. $f1_K(SQN, RAND, AMF) \rightarrow MAC\text{-}A$
4. $f5_K(RAND) \rightarrow AK$
5. $AUTN := (SQN \oplus AK)\|AMF\|MAC\text{-}A$
6. $f2_K(RAND) \rightarrow XRES$; and $f3_K(RAND) \rightarrow CK$; and $f4_K(RAND) \rightarrow IK$
7. $AV := (RAND, XRES, CK, IK, AUTN)$

Figure 2.16 AV construction procedure at AuC.

The UMTS AKA Protocol; The AV Forwarding

The forwarding of the security credentials from the HE to the SN is essentially similar in GSM and UMTS. The MAP protocol, updated for UMTS [58], is still used for transport of the security credentials. The UMTS version of the MAP protocol allows the message exchange to transfer both the GSM triplets and the UMTS AV. The VLR/SGSN may request up to 5 triplets or AVs at a time. The HE may chose to reply with fewer AVs, possibly even with none. It is noted that the MAP protocol messages are in plaintext. An option for protecting the AV transfer exists for the cases when the MAP protocol is transported over an IP connection instead of over the SS7 network. One may then use the so-called SIGTRAN (signalling transport) solution. This possibility has been investigated in TR 29.903 [59]. Security for SIGTRAN is described in RFC 3788 [60].

1. $UE \rightarrow SN$: MM_LOC_UPDA_REQ($IMSI$)
2. $SN \rightarrow HE$: MAP_SEND_AUTH_INFO($IMSI$)
3. $HE \rightarrow SN$: MAP_SEND_AUTH_INFO-ack(AV)
4. $SN \rightarrow UE$: MM_AUTH_REQ($RAND, AUTN$)
5. $UE \rightarrow SN$: MM_AUTH_RESP(RES)

Figure 2.17 The UMTS AKA protocol.

1. **Send** MM_AUTH_REQ(($RAND, AUTN$))
2. **Receive** MM_AUTH_RESP(RES)
3. **If** ($RES = AUTN.XRES$) **goto** SUCCESS
4. **Send** MM_AUTH_REJECT()
5. **Send** MAP_AUTH_FAILURE_REPORT($RAND$) **to** HLR/AuC **and goto** END
6. **SUCCESS: Accept the USIM. Proceed with normal sequence.**
7. **END:**

Figure 2.18 SN action (when USIM responds with MM_AUTH_RESP).

The UMTS AKA Protocol; The Challenge-Response

The UMTS challenge-response procedure is very similar to the GSM challenge-response procedure: The message exchange is identical and the same DTAP [61] messages are used. However, the parameters are not identical. Having obtained one or more AVs from the HLR/AuC the VLR/SGSN may initiate the local AKA procedure by sending the challenge message (AUTH_REQ) that contains the random challenge $RAND$ and the authentication token $AUTN$. The UE verifies that the $AUTN.MAC-A$ is correct for the ($RAND, AUTN$)-challenge and that the challenge is fresh (i.e. not used before or not otherwise expired).

In a successful case the UE will respond with the AUTH_RESP message. The VLR/SGSN then verifies that the response (RES) matches the expected response ($AV.XRES$). The successful procedure is executed in a single round trip (one-pass). The UMTS AKA protocol is outlined in Figure 2.17. The sequence starts with a triggering event, which in the example is an initial registration (the location updating messages have otherwise been omitted). The registration event includes identity presentation with IMSI. The message names have been abridged.

The authoritative reference on the UMTS AKA protocol is TS 33.102 [34], but when it comes to the DTAP message exchange and the message format it is TS24.008 [61] (the stage 3 description) that is the definitive

1. **Receive** MM_AUTH_REQ$((RAND, AUTN))$
2. $f5_K(RAND) \rightarrow AK$
3. $SQN := (AUTN.SQN \oplus AK)$
4. $f1_K(SQN, RAND, AUTN.AMF) \rightarrow XMAC\text{-}A$
5. **If not** $(XMAC\text{-}A = AUTN.MAC\text{-}A)$ **goto** FAIL
6. **If not** $Valid(SQN)$ goto SYNC
7. $f2_K(RAND) \rightarrow XRES$; and $f3_K(RAND) \rightarrow CK$; and $f4_K(RAND) \rightarrow IK$
8. **Send** MM_AUTH_RESP(RES) **and goto** END
9. **FAIL: Send** MM_AUTH_FAILURE(``MAC failure'') **and goto** END
10. **SYNC: Send** MM_AUTH_FAILURE(``Sync failure'',$AUTS$) **and goto** END
11. **END**:

Figure 2.19 USIM action during AKA execution (pseudo-code).

1. **Send** MM_AUTH_REQ$((RAND, AUTN))$
2. **Receive** MM_AUTH_FAILURE(`MAC failure'')
3. **Send** MAP_AUTH_FAILURE_REPORT$(RAND)$ **to** HLR/AuC
4. **Do not accept the USIM. Abort sequence.**

Figure 2.20 SN action during "MAC failure".

source. A reasonably complete description of the UMTS AKA protocol is given in [62, chapters 2 & 3].

Figure 2.19 outlines the procedure at the USIM upon execution of the UMTS AKA protocol. At the SN side there are multiple sequences based upon the USIM reply. When the USIM replies with a normal response the SN acts according to the procedure outlined in Figure 2.18. The UMTS AKA does not have an explicit acceptance message or use any special information element to signal success (Step 6, Figure 2.18). Instead, the continuation of normal signalling subsequent to the AKA procedure implicitly indicates success.

At the SN side one must also deal with USIM rejection and USIM requested re-synchronization. Figures 2.20 and 2.22 outline these procedures at the SN.

1. $SN \rightarrow UE$: AUTH_REQ($RAND, AUTN$)
2. $UE \rightarrow SN$: AUTH_FAILURE(``Sync failure'', $AUTS$)
3. $SN \rightarrow HE$: SEND_AUTH_INFO($IMSI, RAND, AUTS$)
4. $HE \rightarrow SN$: SEND_AUTH_INFO(AV)
5. $SN \rightarrow UE$: AUTH_REQ($RAND, AUTN$)
6. $UE \rightarrow SN$: AUTH_RESP(RES)

Figure 2.21 The UMTS AKA protocol (re-sync).

1. **Send** MM_AUTH_REQ(($RAND, AUTN$))
2. **Receive** MM_AUTH_FAILURE(``Sync failure'', $AUTS$)
3. **Send** MAP_SEND_AUTH_INFO($RAND, AUTS$) **to** HLR/AuC
4. **Await new AVs from HLR/AuC**

Figure 2.22 SN action during "re-synchronization".

The Re-Synchronization Sequence

The quality of the sequence number management is central to the performance of the UMTS AKA protocol. The sequence number management scheme is not standardized, but a set of example schemes is given in annex C in TS 33.102 [34]. The re-synchronization sequence is outlined in Figure 2.21. The sequence starts with an ordinary local challenge-response where the SN already has an AV available.

The information element $AUTS$ is included in the UE response to prove that the re-synchronization was initiated by the USIM and to provide an integrity and privacy-protected indication of the current SQN value at the USIM. The signature $AUTS.MAC-A$ provides the integrity and the use of AK (produced by $f5*$) provides concealment of the SQN_{MS} value. For details of the construction of $AUTS$ see Section 2.6.2 and in particular Equations (2.10), (2.16) and (2.11).

The VLR/SGSN sends an authentication data request (MAP_SEND_AUTH_INFO) with a synchronization failure indication to the HE/AuC. In addition to including $IMSI$ the VLR/SGSN also includes the $RAND$ that triggered the failure and the $AUTS$ parameter as received by the UE. Figure 2.22 depicts the SN action. When the HE/AuC receives an authentication data request with the synchronization failure indication it verifies the $AUTS$ and resets the SQN parameter. Then the HE produces and forwards one or more AVs to the VLR/SGSN. The complete procedure is found in TS 33.102 [34, section 6.3.5].

Formal Verification of the UMTS AKA Protocol

The UMTS AKA protocol has been subjected to formal analysis. Technical report TR 33.902 [63] contains two independent formal validations of the UMTS AKA protocol. The first is found in *Annex A: TLA analysis*. This is a formal analysis of the UMTS AKA protocol by means of the Temporal Logic of Actions (TLA) [64]. The second is found in *Annex B: BAN logic analysis*. Here an enhanced BAN logic scheme, called AUTLOG [65], is used to prove that UMTS AKA meets the required security goals (key freshness in particular). BAN logic was originally defined in [66]. In [62] the Promela language and the SPIN verifier were used to validate the UMTS AKA communications model (challenge-response part). The Promela/SPIN tool [67] is well suited for formal verification of protocol syntax and message exchange. The model in [62] is complementary to the TLA-based model in annex A in TR 33.902.

The UMTS AKA protocol has also been verified with the AVISPA toolset. Here, the UMTS AKA model was expressed in the High-Level Protocol Specification Language (HLPSL, [68]). Interestingly, the HLPSL model only considers "weak authentication". That is, the proven properties only captured authentication without replay protection. To some extent this is a limitation of the model and the modeling tools, but it also reflects the fact that the sequence number mechanisms does not provide full replay protection.

The verifications were all successful. This does not prove the protocol correct, but at least it indicates that none of the more trivial errors are present in the UMTS AKA protocol.

2.6.4 Link Layer Protection

UMTS Link Layer Protection

The link layer protection in UMTS extends from the ME to the RNC (Figure 2.7). This gives a good balance between the security needs and the need to separate access networks from the core networks.

The link layer protection consists of data confidentiality and data integrity services. The data confidentiality service is implemented by means of the stream cipher function $f8$. The $f8$ function is controlled by a 128-bit wide cipher key CK. All user associated control plane and user plane data is protected by the $f8$ function after the encryption commences. However, UMTS permits the operator to use the NULL function (UEA0). This option is only intended for use when local legislation does not permit encryption.

The data integrity service is implemented by means of a message authentication code function ($f9$). The $f9$ function essentially provides message origin authentication and cryptographic integrity protection of the message contents. The scope of the integrity service is user related control plane signalling messages. There is no integrity provision for user plane data in UMTS. The $f9$ function operates under the control of a 128-bit wide key IK and it produces an Integrity Check Value (ICV) of 32 bits. Use of the integrity service is mandatory in UMTS and no NULL function is permitted.

Local Key Distribution

In UMTS the AKA termination points are not the same as termination points for the link layer protection. There is therefore a need for local key distribution.

At the network side the keys CK and IK are stored at the VLR/SGSN and transferred to the RNC when needed. There are several alternatives when it comes to the protocol stack on the Iu-interface (VLR/SGSN–RNC). If the stack is IP-based the operators may use NDS/IP [69] to protect the control plane protocols; otherwise there is no standardized protection scheme for this interface. At the UE side the keys are transferred from the USIM to the ME. There is no mandatory protection scheme for this interface.

Algorithm Negotiations

During setup the UE shall indicate to the network in the *Classmark* information element which cipher and integrity algorithms the UE supports. To avoid bidding down attacks it is necessary that this information is integrity protected. The SN will compare its integrity algorithm capabilities and preferences with those indicated by the UE. If the RNC and the UE have no versions of the UIA algorithm in common then the connection is released. If there is at least one common UIA algorithm then the SN will select one of the common UIA algorithms.

A similar procedure is used for confidentiality algorithm selection. The SN compares its ciphering capabilities and preferences with those indicated by the UE. If the RNC and the UE have no versions of the UEA algorithm in common and the SN/RNC is not prepared to use an unciphered connection (which is permitted), then the connection will be released. If the SN/RNC and the UE have at least one version of the UEA algorithm in common then the SN will select one of the common UEA algorithms for use.

Key Lifetime and the Key Set Identifier

It is not mandatory for the SN to initiate execution of the UMTS AKA protocol for every call setup. The security context lifetime is controlled by a set of counters.[9] The ME/USIM therefore maintains counters (STARTCS and STARTPS) of the bearers that are protected. The counters are compared with the maximum value, THRESHOLD. Should the STARTCS and/or STARTPS reach the maximum value (THRESHOLD) the ME/USIM will invalidate the security context (and thus force AKA execution). The THRESHOLD value is defined by the HE operator and is stored in the USIM.

The key set identifier (KSI) is a is three bit wide number which is associated with the session keys. The KSI is allocated and handled in much the same manner as one did in GSM with the $CKSN$ identifier (see Section 2.3.4). KSI and $CKSN$ also have the same format. The value of "111" is used by the UE to indicate that no valid keys exists. The KSI is allocated by the SN and sent with the authentication request message to the UE where the KSI is associated with the derived CK and IK. The USIM stores separate KSI values for the CS- and PS domain. Use of the key set identifier makes it possible for the SN to identify and re-use the current key set without invoking the authentication procedure.

2.7 Other Aspects of the UMTS Security Architecture

2.7.1 Network Domain Security

The UMTS "Network Domain Security (NDS)" item encompasses basic Core Network security for protection of SS7 and IP based protocols. The scope of NDS is basically aimed at providing network layer protection.

Protection of the MAP protocol

The SS7-based Mobile Application Part (MAP) protocol was originally developed for the GSM system (TS 09.02 [70]). The MAP protocol is also in use in the UMTS core network (TS 29.002 [58]). Amongst the tasks carried out by the MAP protocol is registration of location updating, including transfer of subscriber data, and transfer of subscriber security credentials. Additionally, the commercially important short message service (SMS) data is transported over MAP.

It is clear that the MAP protocol should be secured, but there is a problem. The SS7 protocol stack does not have any standardized security mechanisms. The only available protection for MAP is for the case when MAP is transported over IP instead of over SS7. For this case one may use the SIGTRAN methods (IPsec/TLS). SIGTRAN is briefly discussed in Section 2.6.3.

---

[9] There are separate contexts for the circuit-switched and packet switched domains.

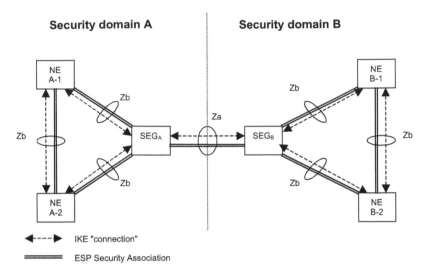

Figure transposed (and corrected) from TS 33.210 (figure 1)

Figure 2.23 The NDS/IP architecture.

Protection of IP Services

For protection of IP-based protocols in the core network it was decided to use native IP security mechanisms and profile them for use within the 3GPP core network. The natural solution was to use IPsec (then RFC 2401) as the basis for the solution. The standardized means of applying IPsec in the UMTS core network is called NDS/IP and is captured in TS 33.210 [69]. Figure 2.23 depicts the NDS/IP architecture. With respect to standard IPsec the NDS/IP solution represents a proper subset, with some strengthening of requirements to guarantee security and to provide better and safer inter-operability. The NDS/IP architecture is primarily designed to protect inter-operator communication, with use of security gateways (SEGs) at the network borders. NDS/IP can also be used within an operator network. The NDS/IP security services are:

- Data Integrity and Data origin authentication.
- Message replay protection.
- Confidentiality (optional).
- Some protection against traffic flow analysis when confidentiality is applied.

To achieve this NDS/IP has standardized on the following:

- *Use of tunnel-mode*
  Tunnel-mode offers full packet-in-packet support and this was deemed
  necessary for NDS/IP. Tunnel-mode is mandatory for use for the
  (inter-operator) Za-interface. Transport-mode *may* be used for the (intra-
  operator) Zb-interface.
- *Use of the Encapsulated Security Payload (ESP) mechanism*
  NDS/IP uses only the ESP protocol. Authentication Header (AH) is *not*
  used at all.
- *Mandatory support for integrity protection*
  In NDS/IP one has mandated use of integrity protection for all NDS/IP
  protected connections. Furthermore, replay protection is made man-
  datory. The weak HMAC-MD5 algorithm is not allowed for use with
  NDS/IP.
- *Confidentiality protection*
  Use of confidentiality protection is not mandatory in NDS/IP, but when
  it is used it shall always be used in conjunction with integrity protection.
  The weak single-DES transform is not permitted with NDS/IP.

The basic NDS/IP architecture relies on the use of pair-wise pre-shared
secrets as the basic long-term security credential [71]. NDS/IP clients can
also use a a PKI-based extension to basic NDS/IP architecture. This extension
is specified in TS 33.310 [72]. The PKI extension can coexist with the basic
pre-shared secret scheme.

2.7.2 Backwards Compatibility Issues

The UMTS security architecture is designed to co-exist with the GSM infrastructure and GSM equipment. Section 6.8 in TS 33.102 [34] details the
various interworking cases. For UMTS subscribers (defined by having a
USIM) the authentication and key agreement will be performed as follows:

- The UMTS AKA protocol shall be executed when the user is attached
  to a UTRAN.
- The UMTS AKA protocol shall be executed when the user is attached
  to a GSM BSS if the user has a ME capable of executing the UMTS
  AKA protocol and if the VLR/SGSN is R99+. This would be the case
  for a user with a UMTS/GSM mobile device connected to a GSM BSS
  attached to a UMTS core network (VLR/SGSN).

The GSM cipher key Kc is derived from the UMTS keys CK and IK by the VLR/SGSN on the network side and by the USIM on the user side. Thus, one has UMTS authentication and GSM key agreement.

- The GSM AKA protocol shall be executed when the user is attached to a GSM BSS in the case where the user has a ME not capable of executing the UMTS AKA protocol.
- The GSM AKA protocol shall be executed when the user is attached to a GSM BSS in the case where the VLR/SGSN is a 2G VLR/SGSN.

Note that the UMTS challenge parameters ($RAND, AUTN$) and the response RES are sent transparently through UTRAN or GSM BSS. The GSM parameters $RAND$ and $SRES$ are also sent transparently through the GSM BSS.

UMTS Subscribers and the Conversion Functions

For the cases where a UMTS subscriber roams to a non-UMTS VLR/SGSN the HLR/AuC shall send a GSM triplet to the VLR/SGSN. The HLR/AuC may then convert an AV to a GSM triplet using the following ($c1, c2, c3$) conversion functions. The conversion function $c1$ (2.21) simply permits the UMTS $RAND$ to be used as-is for GSM.

$$c1 : RAND_{GSM} := RAND \qquad (2.21)$$

$$c2 : SRES := XRES*_1 \oplus XRES*_2 \oplus XRES*_3 \oplus XRES*_4 \qquad (2.22)$$

The $\oplus$ symbol denotes bitwise modulo two addition (XOR). For conversion function $c2$ (2.22) the computation depends on the length of the $XRES$ parameter. The relationship between $XRES$ and $XRES*$ is as follows: The $XRES*$ parameter is 16 octets long and all $XRES*_i$ are 4 octets long ($i \in \{1, \ldots, 4\}$). That is: $XRES* = XRES*_1 \| XRES*_2 \| XRES*_3 \| XRES*_4$. The $XRES*$ parameter is constructed from $XRES$. The $XRES$ parameter has variable length ranging from 32 bits and up to 128 bits in increments of 8 bits. If $XRES$ is shorter that 16 octets then NULL octets are appended until the length of $XRES*$ is 16 octets.

$$c3 : Kc := CK_1 \oplus CK_2 \oplus IK_1 \oplus IK_2 \qquad (2.23)$$

For function $c3$ (Function 2.23) the keys CK and IK are split into halves and then one constructs the key Kc from the CK and IK parts. The GSM-MILENAGE authentication and key agreement functions (TS 55.205 [26]) are based on the use of the MILENAGE algorithm set and the appliance of the conversion functions to derive the GSM triplets.

GSM Subscribers and the Conversion Functions

For backwards compatibility reasons GSM subscribers, defined by having only a SIM available (no USIM), are also permitted to use UMTS MEs in UTRAN networks. For this case one cannot execute the UMTS AKA protocol. The VLR/SGSN will only receive a GSM triplet and so it must run the GSM AKA protocol over the UTRAN network. The GSM AKA protocol can only provide unilateral authentication and it clearly cannot provide UMTS key agreement. For the key agreement the UMTS VLR/SGSN and the UMTS ME shall use the conversion functions $c4$ and $c5$ to derive the keys CK and IK from the basis of the GSM cipher key Kc. The conversion functions are defined in (2.24) (CK) and (2.25) (IK).

The key Kc has at best an entropy of 64 bits and so it is evident that the entropy in the two 128-bit derived keys CK and IK will be very much lower than the normal UMTS level. The practice of converting one 64-bit key into two 128-bit keys is cryptographically highly undesirable, and the operators should only permit use of GSM SIM in UMTS networks for a transition period while deploying UICC/USIM to their customers. Ideally, the keys CK and IK should be individually independent such that the compromise of one key should not automatically lead to compromise of the other key. However, given the limited entropy available one sensibly decided not to split Kc in independent parts.

$$c4 : CK := Kc \| Kc \qquad (2.24)$$

$$c5 : IK := (Kc_1 \oplus Kc_2) \| Kc \| (Kc_1 \oplus Kc_2) \qquad (2.25)$$

The symbol $\|$ denotes concatenation. For Equation (2.25) it is noted that $Kc := Kc_1 \| Kc_2$. As is evident from the function definitions, an intruder with knowledge of Kc or one of the derived keys (CK or IK) can easily compute the other keys.

2.7.3 A Brief Comparison of UMTS and CDMA2000 Access Security

Two of the main 3G architectures to emerge from the IMT-2000 vision are the UMTS architecture specified by 3GPP and the CDMA2000 architecture specified by 3GPP2. The two systems are both based on existing 2G/2.5G systems. Backwards compatibility requirements have meant that these systems have inherited many features and traits from their respective ancestors. The security architecture of the two systems also face many of the same problems, risks, and threats. So, despite the differences between the two systems, it is not too surprising that the access security architectures share a number of common traits and even have a common basis for authentication and key distribution. A comparison of the access security of the two systems is found in [73].

Common AKA Protocol

The UMTS AKA was chosen by 3GPP2 as the basis for the 3GPP2 AKA protocol. The choice was a pragmatic one; the 3GPP2 architecture has many of the same features, traits and backwards compatibility constraints as the 3GPP architecture and so the 3GPP AKA would naturally be a suitable candidate protocol. The actual decision to use the 3GPP AKA protocol was also motivated by the fact that the 3GPP AKA protocol was IPR free.

The UMTS AKA protocol was therefore selected as the basis for CDMA2000 AKA mechanism, with an extension for online verification that the CDMA200 User Identity Module (UIM) is present. The 3GPP2 AKA is defined in [74], which directly references the 3GPP specification (TS 33.102 [34]) for the AKA protocol itself. The CDMA2000 system does not specify use of the MILENAGE algorithm set, but instead specifies its own set of functions in the 3GPP2 specification S.S0055 "Enhanced Cryptographic Algorithms" [49].

Link Layer Protection

For 3GPP2 access security the SN network elements of interest are the Packet Data Serving Node (PDSN), which handles packet-switched traffic, and the circuit-switched nodes VLR/MSC. In 3GPP2 one does not specify privacy and integrity protection for the HE-SN transmission, except to recommend use of IPsec (now RFC 4301 [75]). Tunneling SS7 protocols over IP (SIGTRAN) is also an option for inter-operator transfers.

In CDMA2000 the AES [48] encryption algorithm is used. The data confidentiality protection is based on the IPsec ESP_AES mode-of-operation which is essentially a straightforward counter mode. The AES block size (128 bits) means that in CDMA2000 a special mode-of-operation such as the UMTS $f8$ construction is not necessary. The output of the encryption is used as a stream cipher to encrypt or decrypt the data as required. The counter is initialized to zero at the beginning of each frame, and is incremented for subsequent cipher blocks within that frame. This use is very similar to that of the output of the UMTS $f8$ function.

Similar to UMTS, user data is not integrity protected in CDMA2000. Signalling data sent over the air is always integrity protected. A Message Authentication Code (MAC) is calculated and truncated to the desired size.

The size of the MAC varies depending on the perceived importance of the particular signalling message, but it is always at least 32 bits long. If the message is important in the sense that it provides authentication of the user for some billable service, the generated MAC is called a $UMAC$. The $UMAC$ is computed under control of both the IK and the UAK keys while the "normal" MAC is computed under control of IK only. The UAK key, which is not found in UMTS, is derived by the CDMA2000 $f11$ function and stored on the UIM. The $UMAC$ itself is computed on the UIM. Thus, use of $UMAC$ demonstrates that the R-UIM[10] is still present with the mobile device. The use of UAK is optional. A benefit of the use of $UMAC$ is that it provides for an efficient re-authentication procedure.

In the CDMA2000 system the common broadcast channel periodically distributes a random value *RAND* (not to be confused with the $RAND$ in the AV). When the mobile requests a services it can include the broadcast *RAND* and protect the message with $UMAC$. This proves both the timeliness of the message and the presence of the R-UIM in the handset. For a long-lived data session, it is possible for the network to periodically request an authenticated reply to a random challenge to ensure that the session has not been hijacked.

When a $UMAC$ is required, the computed MAC is passed to the UIM for further processing. A single SHA-1 compression function is evaluated with the UAK as the input chaining value and the input MAC as the data input. The result of this calculation is truncated to the same size as the input MAC and returned as the $UMAC$ value.

Comparison Summary

Both systems are evolved systems and they have inherited features and traits that complicate the analysis. Many non-security trade-offs have been made in the design of the systems that ultimately affects the respective security architectures. The CDMA2000 security architecture has a few improvements on the UMTS security architecture. In particular, the use of function $f11$ to demonstrate the presence of R-UIM is a useful feature. The $f11$ used in CDMA2000 permits efficient re-authentication of the UIM without requiring use of a fresh authentication vector. There are numerous differences between the security architectures, but in practice the systems offer the same level of security. Both can be considered secure within the intended use and both security architectures suffer from being based on a 2G foundation.

---

[10] The UIM can be removable or non-removable. A removable UIM is called a R-UIM and is similar to the UICC/USIM.

2.8 An Analysis of the UMTS Access Security Architecture

2.8.1 High-Level Observations

Several papers have been written on the topic of the UMTS access security architecture. They analyze the architecture to find weaknesses and flaws and many present some minor improvements, but surprisingly few of these paper proposals show a real understanding of the complexity of the system architecture and the underlying trust assumptions, and so on.

What many of the published papers fail to mention is that the by far biggest problems with UMTS security architecture is the restrictions inherited from the GSM 2G architecture. The perception given is that of a 128-bit based security architecture. However, the MAP transport is still almost always unprotected and the UMTS AKA retains the naive delegated authentication approach found in GSM. So, basically, whoever is able to execute a SS7/MAP transaction will be able to retrieve security credentials from the HE. The session keys must also be transported from the VLR/SGSN to the RNC. Some operators protect this link, but many will not (or cannot) due to compatibility reasons, etc. This, of course, puts the whole notion of a 128-bit security architecture in perspective.

Anderson, in his classic paper "Why Cryptosystems Fail" [76], points to the fact many systems had a completely wrong threat model. Now, with UMTS one did in fact have a decent threat analysis [38], but the pressure from backwards compatibility, etc., meant that many of the requirements were silently ignored. Anderson also points to the fact that many security schemes does not fit the system, is inadequate or incomplete. He then asserts that one must take a system level approach to security if one wants to avoid embarrassing shortcomings (like the MAP *AV* transport).

2.8.2 Some Shortcomings of UMTS Access Security

The UMTS security architecture clearly is an improvement over the GSM security. The cryptographic shortcomings have been addressed, but architectural constraints have meant that the security level for UMTS, in several areas, has not improved with respect to GSM. This is problematic since the environment (threats and risks) facing UMTS is much more complex than the simple speech-oriented environment that the GSM system existed in. Some of the problems are listed below:

- **Authentication**
  - *Delegated authentication*

The HE delegates *all* authentication authority to the SN. This must be considered a shortcoming of the UMTS security model.

– *Unauthenticated plaintext transfer of security credentials*
The MAP protocol can only be protected when using the SIGTRAN option.

– *Sequence Number Management*
The sequence number management scheme in UMTS is not standardized. The optional schemes in TS 33.102 [34] depend on proper configuration to be effective.

– *Continued use of SIM-only in UMTS networks*
The functionality for backwards compatibility with GSM means that an operator can deploy and operate a UMTS network with UMTS capable MEs and still use SIM-only smart cards. The authentication provided is then reduced to GSM AKA and the two 128-bit keys (CK, IK) are derived from a single 64-bit key.

- **Key Distribution**

    – *Unprotected transport from the VLR/MSC to the RNC*
    The Iu-interface is a highly complex interface and it consists of both SS7 and IP stacks. For the IP-transport option the NDS/IP [69] scheme may be deployed for protection of the VLR/SGSN–RNC link, but this is not mandatory.

    – *Unprotected transport from the USIM to the ME*
    The original assumption that the USIM–ME interface does not need to be protected is becoming more and more suspect.

- **Rouge Shell**

    – *Compromised ME (Rouge Shell)*
    The new smart phones can be exposed to malware. There is no standardized protection in UMTS against a compromised ME.

- **Link Layer Protection**

    – *Limited integrity protection on the link layer*
    The 32 bit integrity check value is too short for comfort.

    – *Protection Coverage*
    It is problematic that there is no integrity protection for user data.

- **Subscriber Privacy**

    – *Permanent subscriber identity routinely exposed*
    The $IMSI$ is routinely exposed over the Uu-interface.

– *Permanent subscriber identity/location not protected*
An active attacker may page with $IMSI$ and the UE is obliged to answer the paging since there is no way for the UE to determine if the request is valid.
– *Temporary identity may be correlated with permanent identity*
There is no requirement in UMTS on how to generate the $TMSI$. The $TMSI$ may therefore be structured and it may be allocated sequentially. This may make it possible for an adversary to establish the $(IMSI, TMSI)$ association.
– *No requirement on expiry of temporary identity*
The $TMSI$ may be used for a prolonged period.
– *Sequence Number Management*
The SQN may leak information to an intruder since the use of the anonymity key scheme is an option.

2.9 Summary

In this chapter a brief presentation of GSM security has been given. A relatively comprehensive presentation of the UMTS security architecture was given.

The scope and background for the UMTS access security architecture was presented and discussed. It was shown that the security requirements were derived from a set of high level objectives and a threats/risk analysis. The basic UMTS access security architecture is organized around the shared GSM/UMTS identity management scheme, and the UMTS Authentication and Key Agreement protocol is an evolved version of the GSM AKA protocol. The link layer protection was also described, including a presentation of the generic functions ($f8$, $f9$). The chapter ends with a critical review of the UMTS access security architecture. Despite improvements and numerous enhancements many of the fundamental problems with GSM security is still present in the UMTS system. This is not to say that the UMTS systems will be insecure in practice, since most transactions will be adequately protected, but the system is less secure than it could have been.

3

Long Term Evolution

It was the best of times, it was the worst of times, it was the age of
wisdom, it was the age of foolishness, it was the epoch of belief, it
was the epoch of incredulity, ...

– Charles Dickens, The opening lines of "A Tale of Two Cities"

3.1 Introduction

This chapter briefly describes and examines the access security architecture
of the forthcoming LTE system architecture. The LTE architecture is the main
part of Release 8 of the 3GPP and as such it builds on the base of the GSM
and UMTS systems. LTE consists of a new radio system and a new core
network system architecture. Many elements from UMTS has been retained,
but overall LTE is really a new architecture. In particular there are areas where
LTE is significantly different from GSM and UMTS systems:

- LTE is packet-switched only,
- LTE is an All-IP Network (AIPN),
- LTE is not backwards compatible with the GSM SIM smart card,
- LTE provides true mobile broadband services.

The LTE architecture is "packet-switched only" means that LTE does not
support circuit-switched bearer services. You may emulate circuit-switched
services at the network layer or above, but the native bearer level support is
for IP packets only. This seems similar to "LTE is an All-IP Network", but
this also means that only IP based protocol are used, both for the control
plane and the user plane. Thus, the Signalling System no. 7 (SS7) protocols
previously used in the core network in GSM/GPRS and UMTS are now all
gone. Backwards compatibility dictates that the LTE core network will be
able to handle GERAN and UTRAN access networks, and so SS7 may still
be present for those purposes.

An important point is that LTE is not backwards compatible with the GSM SIM smart card or the GSM AKA protocol. To get access to LTE one must be able to run the UMTS AKA protocol. The UMTS AKA protocol is not used directly in LTE, but it is an essential component in the security architecture. The design was quite deliberate; one wanted to be able to use the unmodified Release 99 UICC/USIM in the LTE architecture and one wanted explicitly to prohibit the use of GSM SIM. Needless to say, the break with backwards compatibility with GSM SIM was very important for the security of LTE, since basing a full key hierarchy of 128-bit keys on the very weak basis of one 64-bit key obviously would be very very foolhardy.

The new LTE radio system, E-UTRA, is providing true mobile broadband services. The data rates can easily reach 100 Mbps, but E-UTRA is still not fully compliant with the IMT-Advanced[1] vision of a 4G system. An enhanced version of LTE, called LTE-Advanced, will bring the radio access network up in full compliance with the IMT-Advanced vision of 4G. LTE-Advanced is studied in TR 36.913 "Requirements for Further Advancements for E-UTRA (LTE-Advanced) (Release 8)" [77].

However, the LTE system architecture will not really change with LTE-Advanced and so, from a systems perspective, LTE is the new 4G system architecture from 3GPP.

3.2 Overview over the LTE Architecture

The term LTE started out being used for the new radio system only, but LTE is now the official name (with an associated logo too!) for the new architecture. But the term LTE was not agreed when the work on the specifications started. You may therefore also find that the architecture is sometimes called EPS (Enhanced Packet System) or SAE (System Architecture Evolution). Additionally, the term EPC (Evolved Packet Core) is also occasionally used. Within the security architecture specifications, TS 33.401 [78] and TS 33.402 [79], one will find that SAE, EPS and EPC is used, but rarely is LTE mentioned.

To provide mobile broadband services was a main goal for the new system. From a radio perspective this means that one needs very tight co-

---

[1] International Mobile Telecommunications-Advanced (IMT-Advanced) is the name of an ITU system concept. The ITU IMT concept systems are defining for the cellular/mobile generations and it started out with IMT-2000, which broadly defined the characteristics of 3G systems. IMT-Advanced is in turn defining for the 4G systems characteristics and capabilities.

ordination and control over the radio channels. Thus, the radio termination (the access point, called eNodeB[2] in LTE) must be able to see all data in plaintext. This brought up again the old issue of where in the access network access security should terminate. We have seen that in GSM it terminated in the access point (the BTS). This approach had the considerable drawback that the link from the BTS to the controller (the BSC) was not protected at all. When designing GPRS one wanted to rectify this shortcoming and so one terminated the access security in the core network SGSN node. In retrospect this was a mistake, simultaneously making it harder for the radio system (the quality measurement parameters had to be decrypted in SGSN and sent back to the radio system) and making the core network nodes be aware of access system specific details. So, when designing UMTS one wisely terminated the access link security in the controller (RNC) in the UTRAN. This was a useful compromise, the radio system got its data without too much delay and the core network could ignore RAN specific details.

Two things made the UMTS model obsolete. First, in LTE one has a very clean separation of the control plane signalling and the user plane data. There had been a longtime understanding that the control plane and the user plane had different security needs and furthermore that they could usefully be managed separately. When combined with the dire need for the radio termination (the eNodeB) to see all radio channel data in plaintext it became clear that one needed to terminate security in eNodeB. So, the radio channel and the link layer security then both terminate in the eNodeB. Security requirements then dictate that one re-encrypts data again for the communications further in the network. The control plane and user plane are separated and now terminate at logically disjoint points.[3] Then it no longer made sense to protect these data streams with the same keys, and so one has distinct key sets for control plane and user plane. Therefore it became necessary to design quite an elaborate key hierarchy in LTE security architecture.

3.2.1 The Basic Architecture

A good source of 3GPP architecture level information is TS 23.002 "Network Architecture" [80]. The LTE system model is described in TS 23.401 [81].

---

[2] Also called eNB for short.

[3] The access security part of the control plane terminates at the Access Security Management Entity (ASME). The ASME is part of the Mobility Management Entity (MME). The user plane is forwarded to a Serving-Gateway (SGW) server. User plane security may extend further to the PDN Gateway (in the home network).

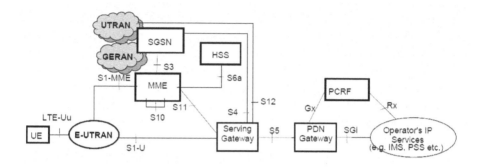

Figure 3.1 LTE system architecture – Non-roaming case.

The document does not only describe LTE, but also how LTE integrates with UMTS and GSM/GPRS. It basically describes two main scenarios, with different configurations, in the specification; the roaming and the non-roaming case. The roaming case includes the home network (Home Public Land Mobile Network – PLMN[4]) and the visited network (Visited PLMN) as separate administrative and operative domains. The non-roaming case is the case where the HPLMN and the VPLMN coincide as depicted in Figure 3.1. The figure is transposed from [81, figure 4.2.1-1]. Security policy considerations may mean that these two cases are handled differently.

The reference model also distinguishes between cases of local breakout (of the user plane traffic) and for cases where the user plane is always tunneled back to the HPLMN. Whether or not local breakout is permitted is an operator choice. In 2G GPRS and UMTS it has been quite common to tunnel all of the user plane traffic back to the home network. From a service quality perspective this has not been a good choice, but it has let the home operator retain full control. In LTE one has the option of local breakout, but with home operator policy control. For the remainder of our discussion we shall mainly look at the roaming case since that is the most generic case. Figure 3.2 depicts the LTE roaming architecture with user plane traffic routed to the HPLMN. The figure is transposed from [81, figure 4.2.2-1].

---

[4] The term PLMN, and the derived Home HPLMN and Visited PLMN, has been in use in ITU recommendations since the ISDN era.

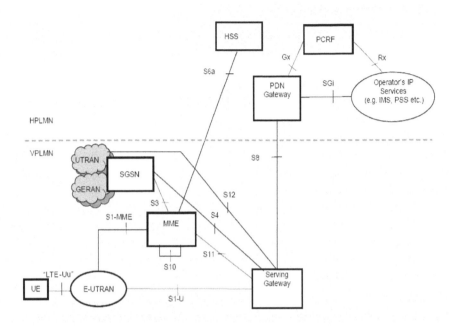

Figure 3.2 LTE system architecture – Roaming case.

3.2.2 The User Equipment

The user/subscriber is still represented by the User Equipment (UE), which minimally consists of the mobile equipment (ME) and the UICC/USIM.

To access LTE the ME must obviously be LTE capable. The UE must also, as a minimum, have a Release 99 compliant UICC/USIM present. A Release 99 UICC/USIM is necessary since one requires the UE to be able to execute the UMTS AKA commands. As seen from the UICC/USIM there is no difference between UMTS authentication and EPS authentication. When it comes to key derivation there are considerable differences and here the UMTS AKA provides only a subset of the EPS-AKA functions. Now, the Release 99 UICC/USIM clearly does not know about the EPS key hierarchy and so it cannot derive those keys, but this is not a problem since the EPS key hierarchy is derived from the CK, IK keys produced by the UMTS AKA protocol. The LTE compliant ME only needs to request the UMTS AKA key pair (CK, IK) and derive the EPS keys.

3.2.3 The Network Elements

The main LTE network elements are the HSS, the MME and Serving Gateway and the PDN Gateway. In addition one obviously also has the eNodeB.

The Home Subscriber Server (HSS)

The HSS is the master database in the HPLMN, and the operative subscription data and security credentials of the subscribers are permanently registered here. The HSS basically fulfills the role that HLR/AuC previously had. A HPLMN may contain one or several HSSs; the actual configuration depends on the number of subscribers, on the capacity of the HSS equipment and on the network organization.

The HSS thus provides support to the call control servers in order to complete the routing/roaming procedures. This includes authentication, authorization, naming/addressing resolution and handling of location dependencies. The HSS is responsible for managing and storing user related information, including:

- User Identification, Numbering and addressing information;
- User Security information: Access control, authentication and authorization;
- Authentication Vectors (AVs): Derived from authentication data;
- User Location information at inter-system level (3GPP-based and non-3GPP based);
- User profile information (subscription service profile, etc.);
- Data related to call control and session management

The HSS uses the Diameter AAA protocol to exchange data with the MME over the S6a interface. The HSS is a control plane entity.

The Mobility Management Entity (MME)

The Mobility Management Entity (MME) contains the Access Security Management Entity (ASME). Logically, the ASME is the network-side terminating point of the EPS-AKA protocol. The MME is therefore an important entity from a security point of view. The main MME functions include:

- Non-Access Stratum (NAS) signalling;
- Non-Access Stratum (NAS) signalling security;
- Authentication and Key Agreement;
- Lawful Interception of signalling traffic;
- Mobility management (registration, etc.);

- Handover handling;
- Roaming (via S6a interface to HSS).

The MME may be implemented in a separate physical node, but it may also be co-located with a Serving-GW in a combined network element. The MME is a control plane entity only.

The Serving Gateway and the PDN Gateway

The two gateways used in LTE are the Serving Gateway (S-GW) and the PDN Gateway (P-GW). The gateways may be implemented in one physical node or in separated physical nodes. The intra-PLMN S5 interface between the S-GW and the P-GW can be realized internally in a combined gateway.

The Serving Gateway (S-GW)

The Serving GW is the gateway which handles user plane traffic towards E-UTRAN. An attached UE is assigned to exactly one S-GW in the VPLMN (at a time). The S-GW is an internal node in the VPLMN and it does not provide external access. The functions of the Serving GW include:

- Local Mobility Anchor point for inter-eNodeB handover;
- Mobility anchoring for inter-3GPP mobility (GERAN/UTRAN);
- Lawful Interception;
- Packet routing and forwarding;
- Quality-of-Service handling, Accounting and Charging functions.

The PDN Gateway (P-GW)

The Packet Data Network (PDN) GW is the gateway which terminates interface towards external IP networks. A UE may access multiple external IP networks simultaneously and so the UE may be connected to multiple PDN GW in parallel. The P-GW functionality is needed for "breakout" and if local breakout is required then the VPLMN must have P-GW functionality. PDN GW functions include:

- Firewall/Per-user based packet filtering (by e.g. deep packet inspection);
- Lawful Interception;
- Packet routing and forwarding;
- Quality-of-Service handling, Accounting and Charging functions.

The eNodeB

The eNodeB is a very important element in LTE. In comparison with BTS and NodeB, the eNB is a fairly autonomous network element. The eNB is

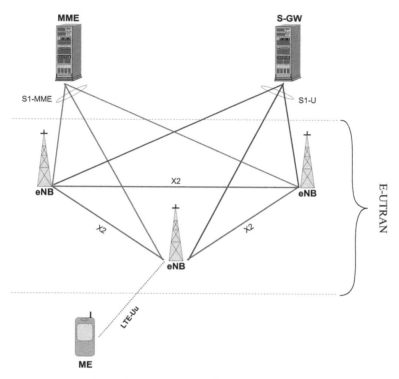

Figure 3.3 E-UTRAN – Overall architecture.

a decision point with respect to many mobility procedures. An eNB may be connected to multiple MMEs and the local eNBs are, or can be, connected in a mesh-like structure. Figure 3.3 shows the E-UTRAN architecture.

TS 23.002 [80] and TS 23.401 [81] provide good overview descriptions of the other LTE nodes, but they do not have a lot to say about the eNB. The eNB and the mobile equipment part of the UE are the radio handling nodes. The radio access architecture (E-UTRA) is specified in the 36-series of technical specifications. Now the actual radio access is a large and important subject, but we shall not go into any detail of the radio access here. We shall only look at the larger picture and then TS 36.300 [82] (detailed overall E-UTRAN description) and TS 36.401 [83] (high-level E-UTRAN architecture description) are good sources.

The EPS-AKA protocol is transparent to the eNB, but it is the receiver of EPS-AKA derived key material from the MME. The eNodeB terminates all link layer security protection from the UE. The connections from the eNB to other eNBs, to the MME or the S-GW are (when required) re-encrypted. We shall have more to say about the eNB security functions in the forthcoming sections.

3.2.4 The Main Interfaces

The following is a subset of all the reference point, protocol stacks and protocols found in the LTE system. All of these have some role to play in the LTE/EPS security architecture.

Stratums

In LTE specifications one frequently uses the term *Access Stratum (AS)* and *Non-Access Stratum (NAS)*. Basically, the Access Stratum is a layer that terminates within the radio access network (RAN). For LTE this means that the AS layer terminates within E-UTRAN; i.e. within the scope of the LTE-Uu and X2 reference points (see Figure 3.3). Correspondingly, the Non-Access Stratum layer is comprised of protocols and data flows that terminates outside the RAN. This applies to traffic towards the MME and traffic towards the S-GW.

Reference Points

The LTE system have a huge number of interfaces and reference points. TS 23.002 [80] and TS 23.401 [81] provide useful overviews over the generic 3GPP architecture and the E-UTRAN access architectures respectively. Most of the interfaces do not concern access security at all and for the sake of brevity and clarity we shall therefore only look at a selected few reference points and interfaces:

- **LTE-Uu (UP/CP):** Between ME and eNB
  This covers the over-the-air interface. Both for user plane and control plane data. The protocols of interest for access security are the PDCP (TS 36.323 [2]), RRC (TS 36.331 [84]) and the NAS (TS 24.301 [85]) protocols.
- **S1-MME (Control Plane):** Between eNB and MME
  S1-MME is a control plane reference point between E-UTRAN and MME. The EPS-AKA protocol runs over this reference point.
- **S1-U (User Plane):** Between eNB and S-GW
  S1-U is user plane reference point between E-UTRAN and the Serving GW. It is used for tunneling of (protected) user data by means of the new

GTP-U protocol.[5] One may protect S1-U with NDS/IP [69] procedures, but it is not required.

- **S6a (Control Plane):** Between HSS and MME
  This reference point is used for transfer of subscription and authentication data for authenticating/authorizing user access to LTE.
- **X2 (UP/CP):** Between eNB nodes
  The X2 interface is split between a user plane (X2-U) and a control plane (X2-C) part.
- **S5 (User Plane):** VPLMN interface between S-GW and P-GW
  The S5 interface is a VPLMN internal interface between the S-GW and the P-GW (see Figure 3.1). It is used when one has "local breakout" for the user plane data. If the S-GW and the P-GW are located at separate physical locations it is recommended that one uses NDS/IP to protect the data. The S-GW and P-GW may also be realized within one physical node and in that case the S5 will be an internal reference point.
- **S8 (UP/CP):** Inter-operator interface between S-GW and P-GW
  The S8 interface is an external interface for inter-PLMN signalling and data exchange (see Figure 3.2). That is, when the S-GW is in the serving network and when the P-GW is in the home network. The user plane data on the S8 interface is transported over the GTP-U [87] protocol.

3.2.5 Control Plane Protocol Stacks

The LTE-Uu and S1-MME Protocol Stacks

The LTE-Uu over-the-air control plane protocol stack and the S1-MME protocol stack is depicted in Figure 3.4. From our viewpoint the NAS, PDCP and RRC protocols are of particular interest.

The S6a Protocol Stack

This interface is functionally similar to the HLR/AuC <--> SGSN/VLR interface in the UMTS architecture, but the SS7-based MAP protocol has now been replaced by the IP-based Diameter protocol. For the S6a interface one uses the base Diameter protocol [88] with the *Diameter Network Access Server Application* (RFC 4005 [89]) extension. TS 29.272 [90] specifies the messages and parameters to be exchanged.

---

[5] For Release 8 the GTP-U protocol is not the version specified in TS 29.060 [86], but the version specified in TS 29.281 [87].

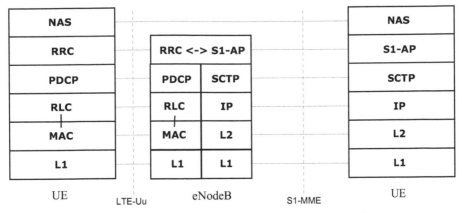

Figure 3.4 The LTE-Uu and S1-MME control plane stacks.

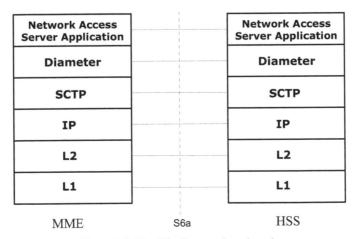

Figure 3.5 The S6a diameter based stack.

The S5/S8 Protocol Stacks

The S5/S8 control plane stack is located between S-GW and P-GW (see Figure 3.2). The GTP-C protocol stack is based UDP/IP (Figure 3.6). The GTP-C protocol is the new GTPv2-C protocol [91]. The protocol may (should) be protected by means of NDS/IP [69]. The S-GW and the P-GW functions may be hosted within the same physical machine or they may be co-located and then one does not need to secure the interface with NDS/IP.

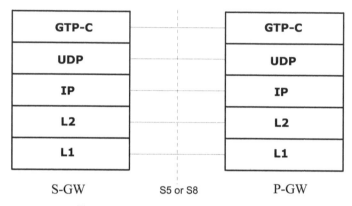

Figure 3.6 The S5/S8 GTP-C based stack.

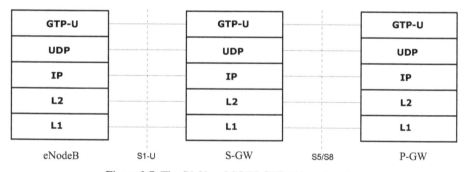

Figure 3.7 The S1-U and S5/S8 GTP-U based stacks.

3.2.6 User Plane Protocol Stacks

The S1-U, S5 and S8 Protocol Stack

The user plane within the E-UTRAN and EPC is all based on the new GTP-U protocol (GTPv1-U [87]). The GTP-U protocol is forwarding the user plane data within the system between the eNodeB and the P-GW. Depending on the operator choice the GTPv1-U protocol may be protected with NDS/IP for one or all of these legs.

3.2.7 The X2 Protocol Stacks

The X2 user plane interface (X2-U) and the X2 control plane interface (X2-C) are defined between eNBs (see Figure 3.8).

The X2 user plane protocol stack is identical to the S1-U protocol stack. The data transfer is by means of the new GTPv1-U protocol [87]. The X2 con-

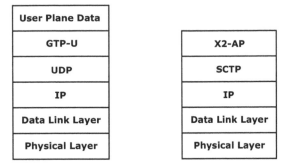

The X2-U (left) and X2-C (right) protocol stacks

Figure 3.8 The X2 protocol stack.

trol plane protocol stack includes the Stream Control Transmission Protocol (SCTP, RFC 4960 [92]) and the X2 Application Protocol (X2-AP).

3.2.8 Essential Protocols

The following protocols are particularly important for access security in LTE.

The Packet Data Convergence Protocol (PDCP)
The PDCP protocol (TS 36.323 [2]) is the transport protocol used between the ME and the eNB. All data confidentiality protection for the user plane is done at the PDCP layer. The PDCP protocol also provides data confidentiality and data integrity protection for the RRC protocol. Data confidentiality and data integrity is briefly described in chapters 5.6 and 5.7 in TS 36.323 [2] respectively.

The Radio Resource Control (RRC) Protocol
The RRC protocol (TS 36.331) [84] handles mobility management and radio resource control for the over-the-air interface. This includes commands to active the security context (Security Mode Command). The data integrity and data confidentiality protection for the RRC protocol itself are provided by the PDCP protocol. However, there are messages in RRC that are not protected. Some because the messages are sent before security is activated. There are also some messages that are sent without integrity protection and some are sent without confidentiality protection (the Paging command is an example). The exempted messages are listed in annex A.6 in TS 36.331

[84]. Data confidentiality protection is recommended, but it is nevertheless an operator option.

The Non-Access Stratum (NAS) Protocol

The NAS protocol (TS 24.301 [85]) handles all control plane signalling between the UE and the MME. In addition to mobility management procedures the NAS protocol also includes the EPS-AKA challenge-response message exchange.[6] Subsequent to the Security Mode Command message all NAS messages are data integrity protected. Confidentially protection is also available as an option. There are some messages that exempted from the protection requirements and those are listed in chapter 4.4 in TS 24.301 [85].

The X2-AP Protocol

The X2-AP protocol defines the radio network layer signalling procedures of the control plane between eNBs in E-UTRAN. The protocol is specified in TS 36.423 [93] and it provides functions such as:

- Mobility Management, including access security functions;
- Load management between eNBs;
- Error reporting;
- Management of X2, including eNB configuration updates.

The access security procedures includes handling of the "X2-handover"[7] where the target needs to obtain key material for processing the connection. This then includes the "chaining" process in which a new K_{eNB} key is derived based on radio context and the old K_{eNB}.

The GTPv2-C Protocol

The GTPv2-C [91] protocol replaces the old pre-Release 8 version of GTP-C (TS 29.060 [86]). Functionally the GTPv2-C protocol carries out the same tasks as its predecessor did, but only GTPv2-C is defined for the new LTE interfaces.

The Diameter protocol (S6a)

Diameter is an Authentication, Authorization and Accounting (AAA) protocol and the base protocol is defined in RFC 3588 [88]. In LTE one no longer uses the old SS7-based MAP protocol for core network mobility management. Instead one uses the Diameter protocol and the Diameter Net-

---

[6] Chapter 4.4 in TS 23.401 [81] contains a useful description of NAS security.

[7] There is also an "S1 handover" type.

work Access Server Application [89]. The way these components are used in LTE is specified in TS 29.272 [90]. In AAA/Diameter parlance TS 29.272 specifies the so-called Attribute-Value-Pairs (AVPs) to be exchanged by the Diameter protocol.[8] The MAP operations for carrying out location management, subscriber data transfer and the authentication procedures (transfer of Authentication Vectors) all have their equivalents in TS 29.272. The message exchange with the `Authentication Information Request` and the `Authentication Information Answer` messages (chapter 5.2.3 "Authentication Procedures" in [90]) is of course of particular interest to us since these are the operations used when the MME requests the EPS Authentication Vector (EPS-AV).[9]

3.2.9 LTE Subscriber Identifiers

The Permanent Subscriber Identity

The venerable IMSI identifier is still with us in LTE as the primary subscriber identity. The IMSI in LTE is defined in TS 23.003 [17] and is the same IMSI as used in the GSM, GPRS and UMTS systems as the permanent subscriber identity. The IMSI consists of the Mobile Country Code (MCC), the Mobile Network Code (MNC) and the Mobile Subscriber Identification Number (MSIN). See Section 2.2 for more information on IMSI.

The Temporary Subscriber Identities

The TMSI temporary subscriber identity used in GSM and UMTS is replaced by a new identity called the GUTI. The Globally Unique Temporary Identity (GUTI) is defined in TS 23.003 [17] and it serves much the same purpose as does the TMSI. Functionally, the GUTI is actually a superset of the the TMSI. The GUTI has following two main components:

- *A unique MME identifier*
  This component, called the Globally Unique MME Identifier (GUMMEI), is constructed from the MCC, MNC and MME Identifier (MMEI). The MCC and MNC have the same format as found in the IMSI.

$$GUMMEI := MCC\|MNC\|MMEI$$

  The MMEI is divided into two subfields, the MME Group ID (MMEGI) and a MME Code (MMEC). The MMEGI is two octets wide and the

---

[8] The AVP is used to encode the operations/commands and their associated data fields.

[9] GSM/GPRS triplets or UMTS AVs may also be requested.

MMEC is one octet wide.

$$MMEI := MMEGI \| MMEC$$

- *A local temporary UE identifier (within the MME scope)*
  The local temporary UE identifier is called the M-TMSI and is 32 bits wide. Thus, the M-TMSI format is exactly identical to the TMSI used in GSM/UMTS.

The GUTI itself then becomes:

$$GUTI := GUMMEI \| M{-}TMSI$$

Within the scope of a single MME, the mobile is uniquely identified by the M-TMSI, but for paging purposes one uses another identifier called the S-TMSI. This permits overlapping paging areas. The S-TMSI is constructed from MMEC and M-TMSI:

$$S{-}TMSI := MMEC \| M{-}TMSI$$

3.2.10 Control Plane and User Plane Separation

In LTE the user plane and control plane are logically separated into separate protocols and messages. The respective planes also have distinct termination points in the network, and while these termination points *may* be at same physical node they are certainly handled by separate logical entities. For instance, the MME is the termination point for most control plane signalling while the corresponding termination point for the user plane (when considering the serving network) is the Serving Gateway (S-GW) (and possibly the PDN Gateway).

The clear distinction between and separation of user plane traffic and control plane data also means that they cannot easily be protected with the same set of session keys. Thus, a consequence of the separation is that the user plane and control plane now needs separate key sets for protecting the respective data. Of course, with separate key set one may also permit different security policies for the planes.

3.3 The Basic LTE/EPS Security Architecture

3.3.1 The Basic Security Architecture

The LTE access security architecture is different from the UMTS access security architecture in a number of ways, but there are also similarities between UMTS access security and LTE access security:

- The USIM is retained unchanged.
- The HSS subscriber security handling is very similar.
- The entity authentication is very similar.
- The Authentication Vector (AV) concept is kept.

The EPS Authentication Vector (EPS AV) is marginally different from the UMTS AV in that the UMTS key set (CK,IK) has been transformed into the K_{ASME}.[10] The K_{ASME} is constructed directly from the (CK,IK) key sets and it is used as the basis for deriving the session keys in the EPS key hierarchy. The HSS knows the difference between EPS AV and UMTS AV. Security contexts derived from EPS AV with the EPS-AKA protocol is termed "Native EPS security contexts".

It is possible to convert between LTE and UMTS security contexts. In LTE parlance a UMTS security context is called a *Legacy security context*. A converted security context is called a *Mapped security context*.

No mapping from GSM security contexts to LTE security contexts is permitted. Thus, a security context created by the GSM AKA protocol *cannot* be used in LTE. This is also true for the case where a GSM security context is mapped to a UMTS security context; this context cannot later be mapped onto a LTE context. This means that subscribers with a GSM SIM cannot access LTE, even when they have a LTE capable terminal. On the other hand a UICC/USIM can be used for access to GSM networks (GERAN) and so it should be clear that the USIM is the only viable choice for operators which plans to deploy LTE access networks.

3.3.2 The User Equipment (UE)

The UICC/USIM

A UE to be used for LTE access must have a LTE capable ME. Furthermore, the UE must have a UICC/USIM subscriber module. A pre-Release 8 UICC/USIM does not of course know about LTE and thus, as seen from the UICC/USIM, the ME is similar to a UTRAN ME. The obvious benefit to this

---

[10] The EPS-AV is then: $EPS-AV = \{RAND, XRES, K_{ASME}, AUTN\}$.

is that any UICC/USIM can be used directly in LTE and that there therefore is no need for issuing and deploying a new subscriber module. The rollout of a new subscriber module to all subscriber may take considerable time for an operator with a large customer base[11] and there is also considerable costs in producing and distributing the subscriber modules. So with respect to time to marked and v.r.t incurred cost, the decision to retain the UICC/USIM is clearly well justified. That notwithstanding, new USIMs (Rel.8) will be able to store the so-called EPS Mobility Management (EMM) parameters. When the USIM supports EMM storage, the security contexts shall only be stored in the USIM.[12] The details of USIM storage are found in TS 31.102 [36] and TS 31.101 [35] describes the UICC.

A consequence of the UICC/USIM being oblivious to the LTE security architecture is that the UICC/USIM clearly cannot produce the keys in the EPS-AKA key hierarchy. So it is clear that the ME must be able to take the output of the UMTS AKA procedure from the UICC/USIM and do the key derivation at the ME. At the UE side it is therefore the ME that derives the session root-key (K_{ASME}) and the other keys in the EPS-AKA key hierarchy.

The Mobile Equipment (ME)

The ME has some new tasks in LTE with respect to security. Amongst others the ME must be able to derive the K_{ASME} root key and the associated derived keys for protection of the user plane, the RRC and the NAS protocols. The ME must also be able to store the keys and their associated security contexts in non-volatile memory in the ME, in case the USIM cannot do so. The ME must also be able to check that the EPS-AKA challenge has the "separation" bit set in the $AUTN$ part of the challenge (see Section 3.3.6).

3.3.3 The eNodeB (eNB)

The eNB access point, in contrast to the NodeB in UTRAN, plays an active part in the access security. The eNB will store key material, it will encrypt and decrypt data and it will also derive key material. So clearly the eNB must be able to process data securely and it must be able to maintain its own physical

---

[11] Expect the process to take in the order of 2 to 5 years.

[12] The USIM stores the data on the UICC, but the definition of the elementary files (EF) to be supported is a property of the USIM.

integrity. Section 5.3 in TS 33.401 [78] gives an overview over the security requirements for the eNB.

eNodeB setup
First of all, it is required that all setup and configuration of the eNBs must be authenticated and authorized. This includes all installation and modification of the eNB software and all parameters that are defined for the eNB.

1. Security associations are required between the EPS core (S1) and the eNB and between adjacent eNBs (via X2). The security associations (SAs) are IPsec SAs according to the NDS/IP [69] requirements.
2. All communication between the remote/local O&M systems and the eNB must be mutually authenticated.
3. The eNB must be able to ensure that software/data change attempts are authorized.
4. The eNB must only use/run authorized data/software.
5. Sensitive parts of the boot-up process must be executed with the help of the secure environment.[13]
6. Confidentiality of software transfer towards the eNB must be ensured.
7. Integrity protection of software transfer towards the eNB must be ensured.

Key Management in eNodeB
The eNB stores and manages subscriber specific session key material. It will also have long-term key material used for authentication of the eNB itself and for security association setup purposes. Therefore it is required that keys stored inside an eNB shall never leave the trusted environment of the eNB, except when explicitly required by a 3GPP specification.

Handling of User Plane Data in eNodeB
The eNB will cipher and decipher user plane packets between the LTE-Uu reference point and the S1/X2 reference points. It is therefore required that user plane data ciphering/deciphering only take place inside the TrE where the related keys are stored. Furthermore, transport of user data over S1-U and X2-U shall be integrity, confidentially and replay-protected from unauthorized parties. However, note that use of cryptographic protection on S1-U and X2-U is an operator's decision.

---

[13] The "secure environment" is generally now called a "Trusted Environment (TrE)" in 3GPP. We will call it a Trusted Environment henceforth.

Generic Requirements for the Trusted Environment

The trusted (secure) environment (TrE) is a hardware based unit logically defined within the eNB. As a minimum it consists of the following:

1. The TrE supports secure storage of sensitive data. This includes all key material and vital configuration data.
2. The TrE support execution of sensitive functions. This includes encryption/decryption of user data, etc.
3. Protected storage of sensitive data. That is, sensitive data used within the TrE shall not be exposed to external entities unless unless explicitly required.
4. The TrE support secure execution of sensitive parts of the eNB boot process.
5. The TrE must be able to maintain its own integrity.
6. All access to the TrE, its data and functions, must be properly authenticated and authorized.

Home eNodeB

In 3GPP one has also defined a "Home eNodeB" (HeNB). The HeNB would be located in people's homes or in other such (generally non-public) locations. The security for HeNB is somewhat stricter and more specific than for the general case for a public-access eNB. Currently the work on this topic has only been published as a *study item* technical report. A study item report has no formal standing and it certainly is not binding for any party, but it is still useful in that it points to security risks , and possible solutions and mitigation strategies for HeNB deployment. The work on eNB is captured in TR 33.820 "Security of H(e)NB" [94].[14]

3.3.4 The Access Security Management Entity (ASME)

The MME/ASME serves roughly the same purpose as would the VLR/MSC and/or the SGSN. The Mobility Management Entity (MME) is a logical entity. Associated with the MME is the Access Security Management Entity (ASME). The MME/ASME are both hosted in a physical server, which may be dedicated to the MME/ASME or which may also contain Serving Gateway (S-GW) and even PDN Gateway (P-GW) functionality. The ASME is the logical contact point for the EPS-AV requests over the S6a interface to the HSS and it is the ASME that handles the EPS-AKA challenge-response procedure

---

[14] Despite the title, the TR also discusses requirements for the (UTRAN) Home NodeB.

towards the UE. In the specifications one often does not distinguish between MME and ASME, and rarely is there a need for a distinction here.

The EPS-AV (see below) does not include the session keys as the UMTS AV did. The ASME must derive the keys of the EPS key hierarchy from the K_{ASME}. The key derivation scheme is described in Section 3.3.9. The ASME also stores all EPS-AKA associated security contexts.

3.3.5 The LTE Compliant Home Subscriber Server (HSS)

The LTE compliant HSS will be aware of LTE and it will distinguish between Authentication Vectors (AV) intended for LTE and AVs intended for UMTS access. The EPS-AV is quite similar to the UMTS AV, but the key pair CK, IK has now been cryptographically transformed into the K_{ASME} master key (see Section 3.3.9). Additionally, the so-called "separation" bit in the $AUTN$ challenge parameter is set to indicate LTE access. The EPS-AV request/reply sequence is carried out by the IP-based Diameter protocol (see Section 3.2.8).

3.3.6 The Authentication Part of the EPS-AKA Protocol

The authentication part of the EPS-AKA protocol is very much based on the authentication part of UMTS AKA. The challenge-response part of the protocol is executed between the USIM and the MME and the AV forwarding is done from the HSS to the MME.

The EPS Authentication Vector

With respect to the AV forwarding there is a minor difference in that the EPS-AV is slightly different from the UMTS AV. The difference is that EPS-AV does not contain any session keys, but instead a root key for the security context. The master key, K_{ASME}, is not used directly as would the CK, IK key-pair that it replaces, but is used as the key material basis for key derivations to construct the EPS session keys. The key derivations, including how K_{ASME} itself is constructed, is detailed in the coming sections. The EPS AV is depicted in Figure 3.9.

The EPS-AV also has one more difference and that is that the AMF part of the $AUTN$ now includes an indication that the AV is an EPS-AV.

AV Forwarding

The AV forwarding takes place between the MME and the HSS over the Diameter based S6a interface (see Section 3.2.8). Figure 3.10 illustrates the

```
EPS Authentication Vector = AV = {
  RAND :       128 bit; -- The random challenge
  XRES : 32--128 bit; -- The expected response
  Kasme:       256 bit; -- The EPS security context master key
  AUTN :       128 bit; -- The authentication token}

Authentication Token = AUTN = {
  SQN  :        48 bit; -- The sequence number
  AMF  :        16 bit; -- The authentication management field
  MAC-A:        64 bit; -- Signature to authenticate the challenge}
```

Figure 3.9 The EPS authentication vector (AV).

Figure 3.10 Distribution of EPS-AV from HE(HSS) to SN(MME).

EPS-AV distribution.[15] The requesting MME in the serving network must specify what type of AV it wants to be returned. This is indicated in the `Network Type` parameter, which is set to "E-UTRAN" for EPS-AKA. The request must also explicitly include the PLMN ID of the requesting MME. The PLMN identifier is specified in TS 23.003 [17] and it consists of a Mobile Country Code (MCC) and a Mobile Network Code (MNC). The HSS constructs the AV according to the requested network type and, for the EPS-AV case, binds the AV to the requesting PLMN ID. For the system to be safe it is required that the `Authentication data request` and the `Authentication data response` messages are authenticated (origin- and message authentication) and confidentiality protected.[16] As of LTE Release 8 there is no direct requirement or procedure in place to secure the S6a interface and the Diameter protocol, but given that Diameter is designed to support

---

[15] Note that the message names used in TS 33.401 [78] and TS 29.272 [90] are not the same.

[16] From a system point of view the `Authentication data request` message does not need data confidentiality, but it is needed for subscriber identity confidentiality reasons.

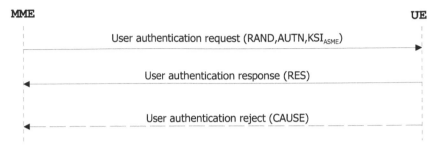

Figure 3.11 EPS authentication.

IPsec or TLS it is of course possible to support the IPsec based NDS/IP [69] framework for protecting S6a/Diameter.

EPS Authentication

The EPS-AKA consists of the very same challenge as for UMTS AKA, with the exception that one now includes a *key set identifier* (eKSI) in the challenge data. There are two possible eKSI types: The KSI_{ASME} or the KSI_{SGSN}. The KSI_{ASME} is used to indicate a native EPS security context while the KSI_{SGSN} is used to indicate a mapped security context.[17] The type information is encoded with the identifer in addition to a 3-bit value field. The eKSI value field follow the same format as the KSI found in UMTS (see Section 2.6.4). The eKSI identifiers are stored with the security context such that one always knows whether the K_{ASME} originated with EPS-AV or UMTS AV.

Another difference is that the "separation" bit (bit 0) in AMF is set to **1** for the EPS-AKA challenge. Previously (before LTE), the AMF was defined to be operator specific, but now the "separation" bit of AMF is reserved to distinguish between EPS-AV and UMTS AV. During authentication over E-UTRAN the ME shall verify that that the AMF "separation" bit is set to **1**. The EPS authentication sequence, see Figure 3.11, is otherwise identical to its UMTS counterpart (See Section 2.6.3).

One thing that is worth noting is that the outcome of the EPS-AKA at the USIM side is exactly as it would be for UMTS. Thus, with respect to the USIM the EPS-AKA and UMTS AKA are identical procedures.

---

[17] These mapped contexts are used during relocation from UTRAN to E-UTRAN: The MME then derives $K_{ASME'}$ from the CK, IK from the UTRAN (via a SGSN).

3.3.7 The Key Hierarchy

Terminology

The following terms are used:

- *Chaining of K_{eNB}*
  Derivation of a new K_{eNB} key based on the existing K_{eNB} during handover.
- *Re-derivation of NAS keys*
  To derive new NAS keys from the same K_{ASME}.
- *Re-fresh of K_{eNB}*
  To derive a new K_{eNB} based on the same K_{ASME} and a freshness parameter.
- *Re-keying*
  To derive a new key (K_{eNB} and NAS keys) based on a new K_{ASME}.
- *Native EPS security context*
  This is a security context based on a K_{ASME} derived by EPS-AKA.
- *Legacy security context*
  This is a security context based on a UMTS AKA security context. That is, the K_{ASME} is based on a key pair (CK, IK) derived by UMTS AKA.
- *Mapped security context*
  A mapped security context is a context created for use in one system and then mapped onto another system. One may map from UMTS to LTE and vice versa.

The Security Contexts

In principle there are three native security contexts types in EPS. First, we have the EPS Security Context. The EPS Security Context consists of the master key K_{ASME} and the associated NAS Security Context and AS Security Context.

Figure 3.12 depicts the relationship between the contexts. As is shown the AS Security Context only exists when the UE is "connected" to the network while the NAS Security Context also exists when the UE is "idle". Generally speaking, an "idle" state means that there are no user plane connection over the LTE-Uu interface.

The EPS Security Context may exist even when the UE is not registered. When the UE registers again the key set identifer ($eKSI$) is used as the reference to the existing EPS Security Context. The EPS Security Context may exist even when the ME is physically turned off. During power-on the ME will retrieve the store context and verify that the originating

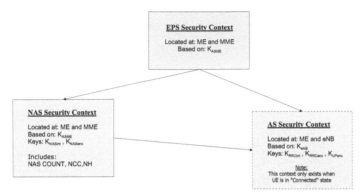

Figure 3.12 EPS context hierarchy.

UICC is still present. If that is not the case then the ME will delete the EPS Security Context. Under no circumstance will the ME permit the K_{ASME} to leave the secure storage of the ME.

A *native* NAS Security Context may also be stored during power-off. If the USIM supports EMM storage the NAS Security Context will be stored in the USIM otherwise it will be stored in the ME.

Master Key for EPS Security Context
In LTE one needs independent key sets for the user plane (UP), the RRC protocol and the NAS protocol (see Figures 3.4 and 3.7). The K_{ASME} master key is 256 bits wide. All derived keys are 256 bits wide, but the session keys are truncated to 128 bits.[18] Figure 3.13 outlines the key hierarchy.

Derived Keys
The following keys are defined:

- K_{eNB} – Root key for AS Security Context
  This key is derived by the ME and the eNB when the UE is connected (ECM-CONNECTED state). The K_{eNB} is a root key for deriving K_{UPenc} and (K_{RRCint}, K_{RRCenc}).
- K_{UPenc} - User Plane
  The K_{UPenc} key is used for data confidentiality protection of the user plane over the LTE-Uu interface.
- K_{RRCint} and K_{RRCenc} – Keys for protection of the RRC protocol

---

[18] There is a requirement for all protocols and interfaces to be able to handle 256-bit wide session keys, but there is no intention of introducing 256-bit wide keys in the immediate future.

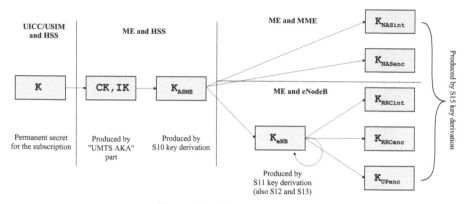

Figure 3.13 EPS key hierarchy.

These keys are derived by the ME and the eNB from the K_{eNB}.

- K_{NASint} and K_{NASenc} – Keys for protection of the NAS protocol
  These keys are derived by the ME and MME directly from the K_{ASME}.

There is additionally a so-called next hop (NH) parameter/key which is used for chaining of AS contexts during handover.

3.3.8 The AS Security Context and Key Derivation

The AS Security Context Root Key K_{eNB}

Before we proceed with the key derivation functions we need to explain a few things about the *AS Security Context* and the root key K_{eNB}.

To start off with, the K_{eNB} key is always associated with a so-called Next hop Chaining Counter (NCC).[19] The "next hop" (NH) refers to a handover specific security parameter, which is actually an intermediate key derived from K_{ASME} (see Section 3.3.9). Associated with NH is the chaining counter NCC. The NCC is incremented for each handover. For the first K_{eNB} derived subsequent to an EPS-AKA event there is no previous handover history and so the NCC is set to zero. At this stage there exists no corresponding NH.

During handover there is a process of chaining in the derivation procedure. The chaining involves producing a new $K_{eNB}*$ key from the intermediate key (NH) or from the current K_{eNB}. Subsequently the $K_{eNB}*$ becomes the current K_{eNB}. There are two distinctive types of K_{eNB} key derivation events:

---

[19] It was ill-advised to call the parameter NCC since that abbreviation is already used for the Network Color Code concept, but within TS 33.401 NCC will only refer to the NH chaining counter.

- **Key Derivation Associated with a handover (*vertical key derivation*)**
  All handover events (intra-eNB, X2-handover or S1-handover) will cause a new K_{eNB} to be generated. This is called *veridical key derivation*.
- **Key Derivation Not associated with a handover (*horizontal key derivation*)**
  When a UE goes from a connected state (ECM-CONNECTED) to an idle state (ECM-IDLE) the eNB will delete all current AS keys, including the NH and the NCC. The ME and the MME will, however, keep the NAS Security Context. Thus the NH and NCC are still present in those entities. If the UE later goes back to ECM-CONNECTED state a new AS Security Context needs to be created. This is involves *horizontal key derivation*.

Model for Key Chaining

Figure 3.14 depicts the key chaining model. The initial state is directly after the EPS-AKA where one has an entirely fresh EPS Security Context and an associated fresh K_{ASME}. A NAS Security Context is also derived at this stage.

Later, when an AS Security Context is needed one will start to derive the initial K_{eNB}. By then there will not be a handover history and the next-hop chaining counter NCC is initialized to zero. The initial K_{eNB} key is bounded to the NAS COUNT (a NAS Security Context message counter). Having derived the initial K_{eNB} one also increments the NCC and derives a NH parameter. This "initial" $(NH, NCC = 1)$ pair is to be stored in the MME and ME, and is used for the initial handover.

Should there be a handover subsequent to this then there will be a *vertical* key derivation. One will then use the "initial" NH as input and derive a new K_{eNB} and new AS session keys. For later handovers the NH will be derived via the previous NH.

Should the UE go to an idle state, the current AS session keys (K_{UPenc}, K_{RRCenc}, K_{RRCint}) will be deleted. When the UE later goes back into a connected mode the K_{eNB} is refreshed and a new set of AS session keys is generated. This is a *horizontal* key derivation event and one then moves one step to the right in Figure 3.14. As can be seen, after the initial K_{eNB} the horizontal key derivation will use the physical cell id (PCI) and $EARFCN-DL$[20] input as for horizontal key derivation. Note that for hori-

---

[20] This is the number of the E-UTRAN specific physical downlink RF frequency used.

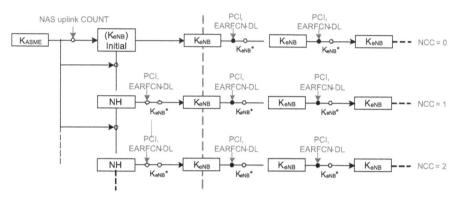

Figure 3.14 Handover key chaining.

zontal key derivation there is also the option of executing the EPS-AKA, in which case one restarts the chaining with a new "initial K_{eNB}.[21]

3.3.9 Key Derivation

In the following we outline the key derivation of the EPS key hierarchy.

The Generic Key Derivation Function

All EPS keys, including the base key material K_{ASME}, are derived by a generic key derivation function (KDF). The KDF is defined in annex A in TS 33.401 [78]. The KDF is inherited from the symmetric-key part (GBA) of the Generic Authentication Architecture (GAA), and is originally specified in annex B in TS 33.220 [95].

The KDF, see (3.1), is based on HMAC-SHA-256. The HMAC construct is specified in RFC 2104 [96] and SHA-256 is specified in NIST FIPS publication 180-2 [97]. This KDF is known as KDF 1 in LTE and as for Release 8 it is the only KDF to be used.

$$derived\_key = HMAC-SHA-256(Key, S) \qquad (3.1)$$

The string S consists of various parameters, depending on which key one is deriving. The general format of S is a bit peculiar. First one has one byte, FC (function code) to distinguish which key derivation one is doing. Then there are **n** parameters encoded as $P\|L$ pairs, where P is the parameter and

---

[21] In particular, the MME may decide that an EPS-AKA should be executed during the ECM-IDLE to ECM-CONNECTED transition.

L is the length (in bytes) of the parameters. The L parameter is itself always two bytes wide. The encoding is peculiar in that it differs from the Type-Length-Value (TLV) coding that is commonly used. The TLV encoding has the advantage that the data is processed such that the length is known before reading the parameter. Thus one can easily handle variable length parameters. In the KDF encoding this is different and the way the encoding is done loses all the advantages of the TLV encoding.

$$S_i = FC\|P0\|L0\|P1\|L1\| \ldots \|Pn\|Ln \tag{3.2}$$

The subscript to the string S, in (3.2), corresponds to the function code (FC).

The K_{ASME} Derivation

The K_{ASME} is derived from a basis of CK, IK. Only the HSS and the UE can derive the K_{ASME}. The derivation is described in annex A.2 in TS 33.401 [78] and the function code (FC) is $0x10$. The "UMTS" keys are concatenated and used as the key to the KDF $(Key = CK\|IK)$. Parameter P0 is a 3 octet encoding of the Public Land Mobile Network (PLMN) ID. The PLMN uniquely identifies the network operator. The KDF specification in TS 33.401 annex A.2 does not explicitly say this, but the PLMN ID in question is, of course, the network identity of the serving network of the MME. Thus, the derived K_{ASME} is cryptographically bounded for use only within a specific serving network.

$$K_{ASME} = KDF(Key, 0x10\|PLMN-ID\|0x0003\|SQN \oplus AK\|0x0006) \tag{3.3}$$

Parameter $P1$ consists of 6 octets and encodes the $SQN \oplus AK$ information element. The $SQN \oplus AK$ element is the masked sequence number, and it is a component of the $AUTN$ parameter. The rules for constructing $SQN \oplus AK$ is found in TS 33.102 [34] (see also section 2.6.3 on page 41 and onwards). The output of Equation (3.3) is a 256-bit binary string.

The Initial K_{eNB} Derivation

This key derivation is generally used for producing K_{eNB} when the UE goes into a connected state (ECM-CONNECTED), but it also applies for idle mode mobility (i.e. when registering within a new area, but without setting up a user plane connection), and for the case when one handovers from UTRAN/GERAN and into E-UTRAN.

The derivation is described in annex A.3 in TS 33.401 [78] and the function code (FC) is $0x11$. The input key is the K_{ASME}. The parameter P0 is set to be the uplink NAS COUNT, which is encoded in four octets ($L0 = 0x0004$). The handling of NAS COUNT is specified in TS 24.301 (Chapter 4.4.3 in [85]). There are individual counters for uplink and downlink and the counters are incremented by one for each NAS message. The NAS messages shall be integrity protected, including replay protection. The actual NAS message in question will vary depending on circumstances; suffice to say that it will be NAS Service Request for theECM-IDLE to ECM-CONNECTED transition and the NAS Tracking Area Update (TAU) Request for idle mode mobility. The inclusion of the NAS COUNT ensures that the AS Security Context is directly bounded to the current NAS Security Context.

$$K_{eNB} = KDF(K_{ASME}, 0x11 \| NAS-COUNT \| 0x004) \tag{3.4}$$

Derivation of the Next Hop intermediate key (NH)
The NH (next hop) is an intermediate key and it is used when one derives chained K_{eNB} keys. The NH is always derived with K_{ASME} as the input key. The derivation is described in annex A.4 in TS 33.401 [78] and the function code (FC) is $0x12$. Parameter $P0$ is the so-called SYNC-input. For the initial NH derivation the SYNC-input will be the newly derived K_{eNB} (the initial K_{eNB}). Subsequently, the SYNC-input will be the current NH.

$$NH_{init} = KDF(K_{ASME}, 0x12 \| K_{eNB} \| 0x0020) \tag{3.5}$$

$$NH_n = KDF(K_{ASME}, 0x12 \| NH_{n-1} \| 0x0020) \tag{3.6}$$

The K_{eNB*} Derivation
This key derivation is for a subsequent K_{eNB}. The input key to derive K_{eNB*} is NH or the current K_{eNB}. Thus, $Key = NH$ or $Key = K_{eNB}$. The derivation is described in annex A.5 in TS 33.401 [78] and the function code (FC) is $0x13$. The parameter $P0$ is set to be the physical cell id (PCI) of the target cell, which is encoded in two octets ($L0 = 0x0002$). Parameter $P1$ identifies the target cell downlink frequency (EARFCN-DL) in two octets ($L1 = 0x0002$). This binds the K_{eNB*} to a specific channel within a specific eNB. When the K_{eNB*} has been derived it will become the new K_{eNB}.

$$K_{eNB} = KDF(Key, 0x13 \| PCI \| 0x0002 \| EARFCN-DL \| 0x0002) \tag{3.7}$$

Derivation of Session Keys
The derivation is described in annex A.7 in TS 33.401 [78] and the function code (FC) is $0x15$. It is used to derive the AS session keys

Table 3.1 Algorithm distinguisher type.

Algorithm Distinguisher	Value
NAS-enc-alg	0x01
NAS-int-alg	0x02
RRC-enc-alg	0x03
RRC-int-alg	0x04
UP-enc-alg	0x05

Table 3.2 Algorithm identifier value.

Algorithm	Identifier	Description
EEA0	0000_2	*Null* ciphering algorithm
128-EEA1	0001_2	Snow-3G based algorithm
128-EEA2	0010_2	AES based algorithm
128-EIA1	0001_2	Snow-3G base algorithm
128-EIA2	0010_2	AES based algorithm

$(K_{UPenc}, K_{RRCint}, K_{UPint})$ from the root key K_{eNB} and to derive the NAS keys (K_{NASint}, K_{NASenc}) from master key K_{ASME}. Equation (3.8) depicts the session-key derivation function.

$$Key = KDF(inKey, 0x15 \| AlgType \| 0x000x01 \| AlgId \| 0x0001) \quad (3.8)$$

The *inKey* is either K_{eNB} or K_{ASME}. The possible algorithm types (*AlgType*) is defined in Table 3.1. The algorithm identities (*AlgId*) are defined in Table 3.2. There are separate identifiers for data confidentiality (EPS Encryption Algorithm (EEA)) and data integrity (EPS Integrity Algorithm (EIA)). Both algorithm identifier types are 4 bits wide. The derived key (*Key*) is 256 bits wide, but only the *n* least significant bits of the *Key* is to be used ($n = 128$ for all session keys in Release 8).

With regard to Table 3.2 it is worth noting that there is no *Null* algorithm for data integrity. The algorithms can otherwise be mixed with the *Algorithm Distinguisher Types* as appropriate, but one would expect all AS keys and NAS keys respectively to be derived from the same cryptographic primitive (Snow-3G or AES).

We can now be satisfied that the problem with missing key/algorithm binding as was present in GSM/GPRS and UMTS has now been solved. In fact, the mode-of-operation weakness that has been reported in the literature has also been resolved with the LTE key derivation scheme. See Section 4.7.4 for a discussion of the algorithm binding problem.

3.3.10 Key Change On-the-Fly

The key sometimes has to change even when there is no idle mode mobility or any handovers. We have two distinct cases which we shall only briefly look into here.

Re-keying

The re-keying procedure takes place after a successful EPS-AKA run. It is always the MME which initiates the EPS-AKA procedure. Since EPS-AKA creates a new EPS Security Context then a corresponding NAS Security Context must be re-derived. The AS Security Context is established only when needed.

AS Key Refresh

Associated with the PDCP protocol there is a 32-bit COUNT parameter. This parameter is constructed from the Hyper Frame Number (HFN) and a PDCP Sequence Number (SN). The PDCP SN can be either 5, 7 or 12 bits long and the HFN occupies the remaining parts of the COUNT parameter. The COUNT is always incremented and the value may, of course, wrap around. When this happens the PDCP COUNT value must be reset. The actual procedure for resolving the issue is based on an intra-eNB handover and from a security point of view the refresh procedure is therefore identical to a handover based key renewal. Thus, the key refresh event is mapped onto a key chaining event.

3.4 Interworking with GERAN and UTRAN

For all the interworking cases the prerequisite is that the UE has a USIM present. Obviously, the ME must also be able to access the different radio access networks.

Under no circumstance shall it be allowed to use a UMTS security context which has been created from a triplet based GSM security context. That is, while the UE is allowed to create CK, IK from Kc (see Section 2.7.2) one shall under no circumstance permit such converted context to be used as the basis for mapping to an EPS security context.

3.4.1 Mapped Context

It is possible to map an EPS Security Context onto a UMTS Security Context and vice versa. Specifically, this means that one must be able to derive K_{ASME} from CK, IK and that one must be able to derive CK, IK from K_{ASME}.

UTRAN to E-UTRAN

The derivation of K_{ASME} from CK, IK is in principle how K_{ASME} is derived anyway. However, there are some significant differences:

- The EPS (CK, IK) never leaves the HSS or the UE, but the UMTS (CK, IK) will always be transferred to the SGSN.
- The MME will receive the UMTS (CK, IK) from an SGSN.
- The MME must be able to create mapped contexts.

This means the MME must also be able to derive K_{ASME}, but the derivations are not the same as for the derivations carried out in a normal EPS-AKA run. In fact, there are distinct derivations for idle mode and during handover (see chapter 9 and annexes A.10 and A.11 in TS 33.401 [78] for all the details). The key derivations are both based on the generic KDF defined in Section 3.3.9 and takes one or two $NONCES$ as freshness input. Function 3.9 shows key derivation of a mapped K_{ASME} during handover while function 3.10 shows key derivation of a mapped K_{ASME} during idle mode mobility. $Key = CK \| IK$.

$$K'_{ASME} = KDF(Key, 0x18 \| NONCE_{MME} \| 0x0004) \qquad (3.9)$$

$$K'_{ASME} = KDF(Key, 0x19 \| NONCE_{UE} \| 0x0004 \| NONCE_{MME} \| 0x0004) \qquad (3.10)$$

E-UTRAN to UTRAN

When the UE enters the UTRAN it needs to have a (CK, IK) key pair. The K_{ASME} is used as the input to a key derivation that is quite similar to the S11 derivation that produced the initial K_{eNB}. In fact, only the function code distinguishes this derivation from the initial K_{eNB} derivation. However, there is a complication in that there is no NAS uplink anymore and so a special NAS downlink token is used as the NAS COUNT. This NAS token is also derived with the generic KDF (see chapter 9 and annex A.9 in TS 33.401 [78]).

$$CK' \| IK' = KDF(K_{ASME}, 0x16 \| NAS-COUNT \| 0x004) \qquad (3.11)$$

The (CK', IK') key pair is derived by the MME (and UE) and is sent from the MME to the SGSN. The SGSN will then forward the keys to the RNC, where they will be used.

An point to be made here is that while the UE and HSS already possesses an EPS (CK, IK) key pair, this key pair will never be used directly in EPS/LTE nor will it be used directly in a mapped context.

3.4.2 Compatibility with GERAN

Basically, as long as the basis for GERAN security is a UMTS security context the rules for mapping contexts will be the same for GERAN as for UTRAN. Chapter 10 in TS 33.401 [78] obviously has more to say on the subject, but in general one derives the keys between E-UTRAN and GERAN as if one had UTRAN as an intermediate step. This means that the UMTS conversion functions are used to create the Kc key from the derived (CK', IK') key pair when one maps an EPS Security Context onto a GSM Security Context. The conversion functions are discussed in Section 2.7.2.

3.4.3 Recommendation

It is not generally recommended to use mapped contexts in E-UTRAN. In fact, one is strongly advised to run the EPS-AKA protocol as soon as possible after entering E-UTRAN. If one is re-entering E-UTRAN one may attempt to re-active a stored EPS Security Context. There are no recommendations against using an EPS context mapped to UTRAN or GERAN.

3.5 The EPS Algorithms

The EPS algorithms outlined in Section 3.3.9 (Table 3.2) are fully defined in annex B in TS 33.401 [78].

3.5.1 Data Confidentiality

One has two basic cryptographic primitives, namely Snow-3G [53] and AES in counter mode (CTR) [98]. The EEA mode-of-operation in LTE is exactly identical to its UMTS cousin. When using Snow-3G there is no difference between Snow-3G in UMTS and Snow-3G in LTE. For use of AES CTR one needs to specify the initial 128-bit T parameter (counter block), and this is done assigning the COUNT, BEARER and DIRECTION information to T. This only amounts to 32, 5 and 1 bits respectively. The remaining T bits are all set to zero. Subsequent T blocks are obtained by standard incrementing functions defined for AES CTR.

Both EEA1 (Snow-3G) and EEA2 (AES CTR) are set up according to the UMTS $f8$ mode-of-operation (see Figure 2.11). The only difference is that the name is EEA instead of $f8$ and that the key CK is replaced by the generic KEY. The KEY in question can be either the K_{UPenc}, the K_{NASenc} or the K_{RRCenc}.

3.5.2 Data Integrity

One has again the same two basic cryptographic primitives, namely Snow-3G [53] and AES. For AES one uses the CMAC mode [99]. The EIA mode-of-operation in LTE is almost identical to its UMTS cousin, but with FRESH being replaced with BEARER. BEARER is only 5 bits long and so the remaining 27 bits are all set to zero.

For use of AES CMAC one needs to map and adjust the LTE parameters to the CMAC requirements. Thus the length of MESSAGE becomes the BLENGTH parameter in CMAC. The first 64 bits of the M (message) parameter of CMAC is assigned values from COUNT (32 bits), BEARER (5 bits) and DIRECTION (1 bit) and padded with zeros up to 64 bits. The MESSAGE itself is then appended to M such that:

$$M = COUNT\|BEARER\|DIRECTION\|(26-zeroes)\|MESSAGE$$

Correspondingly, $Mlen$ becomes $Mlen = BLENGHT + 64$. The MACT (MAC output) has length Tlen, which is set to 32 bits ($Tlen = 32$).

Both EEA1 (Snow-3G) and EEA2 (AES CMAC) are set up according to the UMTS $f9$ mode-of-operation (see Figure 2.13). The differences are minimal: the name is EIA instead of $f9$, the key IK is replaced the generic KEY, the COUNT-I is named to COUNT and the FRESH has been replaced with BEARER-ID. The KEY in question can be either the K_{NASint} or the K_{RRCint}.

3.6 Summary

In this chapter we have briefly investigated the LTE security architecture. We saw that for the subscriber the USIM is the basis for all LTE/EPS security, but that the key hierarchy in LTE also required the ME to derive keys from the UMTS AKA output. The impact of LTE on the network side is quite substantial. The EPS security architecture here mirrors the general architecture and it reflects the decision to separate the user plane and the control and the decision to terminate (most of) the access security data confidentiality and data integrity in the eNodeB access point.

The presentation of the EPS/LTE security architecture has been amended and abridged from what you will find in the EPS security architecture specification (TS 33.401 [78]). This, it is hoped, will improve clarity and allow the reader to get a grasp of the concepts, if not the details (of which there are a lot!).

It must also be mentioned that the EPS/LTE security architecture specification has a companion specification, TS 33.402 "Security aspects of non-3GPP accesses" [79] which has not been addressed. This specification details how to do access security for non-3GPP based radio systems. This is in fact an important area since one may easily hook up new radio systems to the EPS core network (EPC). This means that 3GPP2 based radio access and IEEE based WiMAX may be attached to a 3GPP EPS core network and that one may base this on access security as specified in TS 33.402. Commercially, this makes a lot of sense. A 3GPP operator may then leverage the existing infrastructure while still being able to supplement its radio coverage with complementary radio access technologies. Even for "fixed" xDSL replacement technologies this approach is useful.

It cannot be denied that the access security architecture in LTE is substantially more complex than its UMTS cousin. Still, it is hoped that the reader has obtained a useful introduction to access security in a real-world mobile broadband system and understands the basic ideas behind the EPS security architecture.

4

Access Security for Future Mobile Systems

> The city's central computer told you?
> R2D2, you know better than to trust a strange computer.
>
> *– C3PO (Star Wars, Episode V: The Empire Strikes Back)*

4.1 Introduction

In this chapter various aspects of a possible future mobile access security architecture is described and examined. The objective is to derive assumptions and requirements for a Privacy Enhanced 3-Way Authentication and Key Agreement (PE3WAKA) protocol. The subscriber privacy requirements are discussed in Chapter 5.

The discussion will take into account aspects of the target system architecture, cellular environment design constraints and system performance requirements. This includes an analysis of the requirements of the three principal entities: the UE, the HE and the SN. A subsection is dedicated to discussing the possible trust relationships between the principals. The assumptions behind the trust relationships should be mapped onto the control model and enforced by the security architecture.

4.2 Mobility Management Aspects

4.2.1 Mobility Management

Cellular systems aim at presenting the subscriber with a managed link layer that appears to the subscriber as a large single link segment. This does not preclude further mobility management handling at higher layers. In environments with incompatible RANs under control of a single SN, network layer handover management is needed if handover functionality is required across the access radio access systems.

4.2.2 Location Registration

The *location registration* procedure includes the attachment to an SN server, the updating of location information and more at the HE subscriber register and the transfer of subscription information to an SN server. Thus, the location registration procedure establishes the basic mobility context. The *location updating* procedure is a renewal procedure in which the preexisting mobility context is renewed and/or updated to reflect changes at the UE (location change; may include other changes to the UE status). A *cancel location* procedure would de-register the mobility context.

In the GSM/UMTS location updating scheme the *security context* establishment has been separated from the *mobility context*. This is unfortunate since the creation of the two contexts is essentially triggered by the same external event (UE movement into a new SN server area or expiry of the existing context).

4.2.3 Handover and Cell Reselection

Handover and Control Model

The handover procedure(s) should be closely correlated with the control model. The control model used in the current public cellular networks is basically a network-centric control model. In the 3GPP systems we have the following:

- *Location handling:* The location handling is principally controlled by the HE for the global location management. Local location management is handled by the SN.
- *Handover:* The handover decision point is located in the SN. The controlling node should have intimate knowledge of radio link quality and load conditions.
- *Security Context:* The security context is established by the AKA protocol. In the 3GPP systems the AKA protocol is executed in two stages, and effectively the AKA protocol delegates the identity corroboration procedure from the HE to the SN.

  The UE is only a responder in the 3GPP AKA protocols. In GSM AKA protocol the MS(SIM) is left without any control or authority. In UMTS AKA protocols the UE(USIM) verifies the challenge and it rejects the SN if verification fails. The EPS-AKA protocol is similar to the UMTS AKA protocol in this respect.

- *Link Layer Protection:* The link layer protection covers parts of the $(UE \rightleftarrows SN)$ channel. In the 3GPP systems the SN is always the initiator of link layer protection.

Handover Decision

The algorithms for deciding when to execute the handover procedure is essential to the mobile system. The handover strategy needs to be consistent with the service model of the system. In public cellular systems one tends to favor a network centric control model, and it is assumed that this will also be the case for future mobile systems. There are important technical arguments for a centralized control model. In particular, the AN/SN will have a better overview of the resource utilization of the present radio environment than the user device can have. It is nevertheless assumed that the mobile device will assist the network in preparing for handover events.

Handover and Security Context Handling

The main aspect to consider is the need for security context relocation and/or security context reestablishment. The approach taken here is that one should not generally relocate the session security context. That is, the preferred solution is to re-run the AKA protocol or run a dedicated rekeying procedure instead of relocating the security context to a new node.

Normally there is a security context anchor point in the AN/CN. In GSM/UMTS the anchor would be the VLR/SGSN and in LTE it would be the MME. A handover that causes a change in the security context anchor point should invoke the AKA protocol. Other changes may also trigger AKA execution, including moving between AN controllers and moving between AN types (i.e. different radio system technologies). This implies that a heterogeneous handover procedure should invoke the AKA protocol.

4.3 System Model

4.3.1 Architecture Aspects

An important part of a future access network is the security architecture. The requirements on the security architecture would permit the HE, SN and UE to establish security contexts and to update the contexts as the UE moves between ANs. The security context management should therefore be independent of any particular (R)AN. However, when it comes to access link protection between the UE and the RAN there will likely be RAN specific solutions. The access protection should normally be provided at the link layer. The protection must then be adapted to the specifics of the target link layer. The requirements for the protection should nonetheless be quite similar. Basically, one needs data integrity protection[1] and data confidentiality protection.

In this section a set of new nodes is proposed. To a large extent, the node functionality is already present or suggested for the 3G and beyond-3GPP system architectures. The reason for not using the 3G/beyond-3GPP node names is to distinguish them from the particulars of the current/suggested nodes and to instead define new (semantically neutral) names.

4.3.2 User Device

The UD contains the User Entity (UE) and at least one Mobile Termination (MT) unit. The UD and UE will be discussed further in Section 4.5.2. The UE will include a Secure Module (SM) on which the long-term security credentials are stored. The SM is discussed further in Sections 4.5.2 and 4.8.2.

The SM is tamper-resistant and compromise is a very infrequent event. The remainder of the UD should contain procedures and mechanisms that provide a certain level of system integrity. UD compromise (excluding the SM) is expected to occur relatively infrequently, but nevertheless more often than SM compromise. The HE must be able to routinely handle compromised UDs and this calls for remote device management.

4.3.3 Cell and Cell Area

The *cell* concept is assumed to be kept. The cell is implemented by an Access Point (AP) node. One AP may serve multiple cells. Each AP will have an assigned address. Each individual cell will have a unique address/identity. The *Cell ID* may include the AP address as a subfield. The Cell Area is the service coverage area of a cell.

The APs will by their very nature be highly distributed. The AP will thus be exposed to both physical and logical attacks. From a system perspective one should expect that APs may occasionally become compromised. Detection of AP compromise will therefore be essential and the AN must additionally be able to routinely handle AP compromise incidents.

---

[1] In 3GPP terminology this would include data origin authentication and data integrity protection.

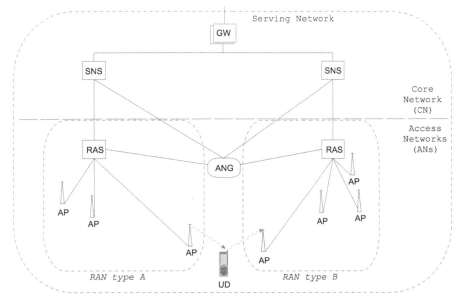

Figure 4.1 Cellular architecture: Inter-RAN communication.

4.3.4 Radio Access Server and Corresponding Location Area

The RAS node is a node somewhat similar in functionality to the BSC/RNC. It will have one or more assigned addresses. The RAS will administer one or more Location Areas (LA). Each LA will have a unique identity. The LA identity is published on a cell broadcast channel as is the case for the present-day 3GPP systems. The LA, which consists of set of cells, has the same role as in the 3GPP systems.

The number of RAS nodes is considerably smaller than the number of APs and the RAS nodes are considerably more valuable. It is therefore reasonable to expect that they will have better physical protection. The RAS nodes are therefore better suited as the network-side security termination points than the APs. However, in LTE we saw that radio management requirements meant that the eNodeB (the LTE AP) had to see the transmitted data in clear. Thus, there is still a convincing case for terminating link layer security in the AP.

4.3.5 Radio Access Network

The RAN consists of a set of RAS nodes and links, with corresponding underlying AP/Cell structure, such that they together form one virtual homogenous link layer segment. The assumption is that handovers can be seamlessly performed within the RAN. The RAN may need assistance from core network servers to carry out inter-RAS handovers, but inter-RAS HO functionality may also be located in the RAN. In a future mobile system a SN will likely control multiple RANs with different technological bases. In a heterogeneous link layer environment it will not always be possible to carry out link layer handovers. Fast inter-RAN handovers may still be desirable and so one ought to optimize the RAN interconnections to allow this. To facilitate this one may need to deploy Access Network Gateway (ANG) nodes deep into the RANs. These ANGs would cross connect RANs and coordinate inter-RAN handovers. Figure 4.1 depicts an inter-RAN scenario where the ANG also has connections with the Serving Network Server (SNS). The ANG may then play the role of an AN anchor. The ANG need not be a physical node; it may be a functional entity integrated with the RAS servers.

All RAN communication is expected to be run over authenticated and protected links. With respect to inter-RAN handovers it is desirable if the ANG is able to handle security context re-establishment. The ANG would then have some similarity with the MME found in the LTE architecture. The ANG may be the node of choice for the network-side AKA termination and the RAS may be the node of choice for termination of the access link protection.

4.3.6 Serving Network Domain

The SN will consist of servers (Serving Network Servers (SNS)) which interface with the RANs. A core network would interface both cellular ANs, wireless ANs (with only rudimentary mobility handling) and fixed line ANs. The interworking level will likely vary, but as a minimum all these ANs would provide an IP-based network layer.

The SNS should not need to know the specifics of the physical radio link. A SN will have one or more Gateways (GWs). The GWs will interface with the external networks, including HE networks and other SN network. The SNS will be similar in functionality to the VLR/MSC and the SGSN. The GWs will define the gateways of the SN. The SNS may be the SN AKA termination point. If so, the session keys (session context) must be securely forwarded to the handling RAS. The SN nodes are assumed to have good physical protection. All communication within the SN is assumed to be protected and all GW communication with external networks is assumed to be protected. The SNS–RAS/ANG communication is similarly assumed to be protected.

4.3.7 Home Entity Domain

The HE in a future mobile system will functionally be relatively similar to the current HEs.[2]

The HE will consist of a set of subscriber databases (Home Registers (HRs)). The HRs will together serve all of the HE subscribers. An HE will additionally have one or more GWs. The GWs will interface with the external networks, including other HE networks, SN networks, third party networks, etc. The HR will also be the HE AKA termination point. The HR also keeps and maintains all long-term UE security credentials. The HE nodes are assumed to have good physical protection. All CN communication within the HE is assumed to be protected. The GW communication with external networks is also assumed to be protected.

A network neutral security context setup in an environment with heterogeneous ANs would require that the AKA protocol either be technology neutral with respect to the lower layers or that separate adaptations of the protocol is made for each AN. With the prevalence of IP-based services it seems overwhelmingly likely that future systems would have a strong focus on IP provisioning. It therefore seems likely that the security context setup protocol will be designed to run over the IP stack. Given the fact that the AKA protocol would have a variety of ANs it would seem best to let the AKA produce generic key material (this is the case in LTE). That is, the key material derived through the AKA process should be independent of the particular AN. Subsequent to the AKA execution one should derive local AN adapted session key material from the AKA key material.

4.4 Simplified Communication Model

4.4.1 User Plane and Control Plane

The terms *User Plane* and *Control Plane* is in common use in formal standards (3GPP, ETSI, ITU-T). The distinction is useful and the term is also used in the present document. The terms denote the following:

---

[2] In the 3G case HE stands for Home Environment. For an imaginary future system HE stands for Home Entity.

DEFINITION 1 (USER PLANE). *The User Plane consists of traffic/data that originates or terminates with the user/subscriber.*

DEFINITION 2 (CONTROL PLANE). *The Control Plane consists of system signalling data.*

The Control Plane can be associated with a user/subscriber session or it can be independent of any particular user/subscriber.

The definition of the *user plane* may include control plane information for higher layer signalling. That is, the definition of *user plane* and *control plane* are relative to the defining layer. For our purpose the defining layer is the link layer at the A-interface (Figure 4.2). The distinction may be less clear for the B-interface, but all mobility management and security management signalling will be considered to be on the control plane.

4.4.2 Communication Model: Paths and Interfaces

To simplify the design and analysis of the access security architecture a simplified communication model is needed. This model does not account for the complexities of the RAN interconnections and ignores the problems associated with UE–SN protection termination point and the AKA termination points.

Entity Interfaces

The principal entities can only communicate with each other over the network as defined in Figure 4.2, which depicts a simplified topological model of the communication paths between the principal entities. It is noted that this communication model is for *control plane* signalling. Subscriber originated *user plane* traffic may exit/enter the mobile domain at the SN. As a matter of notation we denote A and B to be termination points and to let $(A \rightleftarrows B)$ denote a bi-directional communication path between A and B. Then we have:

- **A-interface (UE–SN)**
  The A-Interface defines the $(UE \rightleftarrows SN)$ path over the radio access network interfaces between the UE and the core network part of the SN. Generally, the A-interface will be the part of the communications link that restricts the offered bandwidth and it will also be the interface with potentially the most severe error conditions, longest delays and worst jitter conditions.

- **B-Interface (SN–HE)**

  The B-interface defines the $(SN \rightleftarrows HE)$ path over the interface between the core network of SN and the HE. The offered bandwidth and the quality is assumed to be sufficient and predictable. The propagation delay is expected to be fixed.

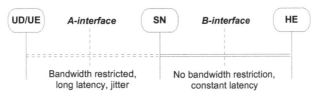

Figure 4.2 Control plane interfaces and communication paths.

The $(UE \rightleftarrows HE)$ path is composed of the $(UE \rightleftarrows SN)$ and $(SN \rightleftarrows HE)$ paths. It is noted that all UE-HE communication *must* pass via the SN. The paths also define the possible channels: CA, CB and CC. The channels are equivalent to the following paths:

$$CA \equiv (UE \rightleftarrows SN)$$

$$CB \equiv (SN \rightleftarrows HE)$$

$$CC \equiv (UE \rightleftarrows HE)$$

4.4.3 Security Assumption for the A-Interface

All user-session associated control plane communication is expected to be protected (integrity and confidentiality). All user plane data transferred over the A-interface (the CA channel) is expected to be protected (integrity and confidentiality).

System broadcast information is on the control plane. The broadcast information is inevitably meant to be readable by all legitimate users. Commonly this information is in plaintext and not integrity protected. The same is true for dedicated broadcast information like *paging* messages that in the current 3GPP systems are transmitted unprotected and in plaintext. Similarly, some of messages exchanged during context establishment will be in unprotected plaintext. All unprotected messages must initially be treated as unverified claims, and should be corroborated at a later stage.

4.4.4 Security Assumption for the B-interface

The control plane communication between the HE and the SN will consist of subscriber specific signalling and signalling that is independent on any specific subscriber activity. There is therefore a generic requirement for the HE and SN to secure the CB channel. The generic protection of the CB channel should be established in conjunction with the establishment of the channel itself. One must therefore require the CB channel to be authenticated and it seems reasonable to assume that both confidentiality and integrity services are needed.

User Plane traffic over the B-interface may also need protection. This could be provided as an integral part of the core network services. It could be provided end-to-end and would then not be part of the access security procedures.

4.4.5 The CC Channel

For the purpose of remote security management of the UD and the SM it may be useful to establish a dedicated protected CC channel. This channel is an end-to-end channel between the UE and the HE, and it would be part of the AKA context.

4.4.6 Actor Reference Points and Intruder Interception Points

In this subsection the notion of *regulatory actors* and an *intruder* is introduced. These concepts are further described in Sections 4.5 and 4.10 respectively. The regulatory actors only operate on fixed and well defined reference points.

- R_{SN} – A regulatory interception point within the SN domain.
- R_{HE} – A regulatory interception point within the HE domain.

The intruder will not be so polite as to use fixed reference points; it may in principle attack wherever it wants. Still, in the given communication model there are a limited number of places were an intruder can execute attacks. As depicted in Figure 4.3 the intruder can in principle only attack at the A- and B-interfaces and the external exit/entry points (called U_{SN} and U_{HE} here). The intruder depicted in Figure 4.3 is the standard Dolev–Yao Intruder (DYI) [100], which can intercept and modify all communication, but which does not physically corrupt the principal entities. The DYI may also attempt to masquerade as a legitimate principal or actor. The DYI may of course also

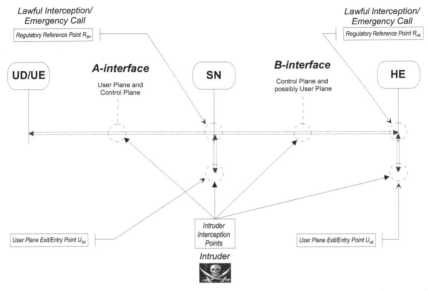

Figure 4.3 Regulatory reference points and intruder interception points; the intruder may also attempt to intercept the regulatory reference points.

be a legitimate principal or actor, trying to gain unauthorized access to some resource.

The DYI will in principle get access to all data transferred over the both the CA and CB channels. This would be all user plane data and all control plane data, including non-user related control plane data. While the scope of the access security procedures clearly is limited to the system access, it must be noted that an intruder that monitors user data in encrypted/protected form at the A/B-interface and in unprotected form outside the mobile system will be able to gather plaintext/ciphertext pairs. This may facilitate certain cryptographic attacks.

4.5 Principals and Regulatory Actors

In order to discuss and analyze security requirements we need to introduce some definitions.

4.5.1 Principal Entities

To simplify our analysis we only define three principal entities.

- **User Entity (UE)**
  The UE is a subscriber entity, and the subscription is registered with the HE. The HE has security jurisdiction over the UE. The HE assigns the permanent HE-UE security credentials and unilaterally decides on the permanent UE identity ($UEID$).
- **Home Entity (HE)**
  The HE is the *home operator*. It is responsible for service provision to the UE. It manages macro mobility and handles charging and billing on behalf of the UE when it consumes access services.
- **Serving Network (SN)**
  The SN provides a core network (CN) with a non-empty set of access networks (ANs) attached to it. Through HE-SN *roaming agreements*, the SN provides access services to UEs. The roaming agreements are mutual in nature.

It is common for cellular operators to own and operate both HE and SN services. Virtual operators will only have HE functionality. One should seek to minimize the trust dependency between the principals since one cannot rule out that a legitimate principal, by accident or by intent, misbehaves. The aim should be to prevent and/or reduce the effect of a misbehaving or compromised principal.

4.5.2 The User Device

User Device Composition

The User Device (UD) can include a powerful general purpose computing platform or it may be a dedicated device with scarce resource. As depicted in Figure 4.4 a UD consists of one or more mobile termination (MT) units and one or more Secure Module (SM) units, a clock (CLK) and positioning device (PD) and the computing platform.

A SM is owned/controlled by exactly one administrative entity, the HE operator. Each SM contains one or more subscriptions, and we shall assume that all subscriptions on a single SM are issued by the same controlling HE. The SM may be implemented as a smart card, as is common in 3GPP systems, but the actual type of implementation is not important. However, we do require the SM to provide secure storage and a computation engine for processing cryptographic primitives and protocols. This means that the SM must have hardware support (i.e. no software-only solution).

Each user subscription is represented by exactly one User Entity (UE). The UE, which is an application running on the SM, is uniquely referenced by

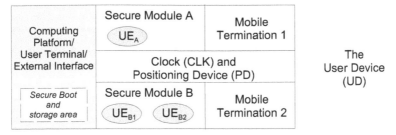

Figure 4.4 The user device.

the UE identity ($UEID$). Figure 4.4 illustrates the possibilities. We normally refer only to the UE, but sometimes one implicitly includes SM and/or MT capabilities in the discussions.

All parts of the UD must to some extent be trustworthy and they must be able to communicate securely.

Risks Associated with the User Device
There are many risks to the UD/UE, including the following main risks:

- *The UD may be stolen*
  In addition to the loss of the UD itself the subscriber risks losing content on the device, including contact lists, pictures, films and music. The subscriber may also risk that the UD is used to make calls and consume paid services in the networks.
- *The UD may be compromised*
  The UD may be compromised in many ways. If malware like trojans, computer viruses, etc., takes over the device the malware may also be able to gain control over the authenticated communication channels.
- *The UD may be lost*
  A misplaced device may be lost entirely, but the subscriber may not know whether the device was lost or if it was stolen (or maybe found by an opportunistic human intruder). The risks are similar to the case of a stolen device unless the user knows that the device is permanently lost.
- *The UD may be destroyed*
  If the user/subscriber knows that the device has been destroyed then the risk is mainly that of losing a valuable device and that of losing the device contents.

It is noted that access security schemes alone cannot prevent the above risks.

The User Entity
The features of the UE include a globally unique UE identity ($UEID$) and the means to authenticate the $UEID$. Physically, the UE must be located such that it can protect its own integrity and provide confidentiality for secret keys, etc. In GSM, UMTS and LTE this was solved with placing the subscriber credentials on a smart card. In CDMA2000 the subscriber credentials are either placed on a removable (R-UIM) or intergrated with the handset (standard UIM).

The Secure Module
The Secure Module is assumed to be hosted on a tamper resistant device. The tamper resistant device may be a stand alone smart card or it may be integrated with other UD components. The SM will execute the UE application, which contains sensitive information. If the SM integrity is broken, which may happen, then the SM should ideally detect this and disable itself. The SM must be resilient against physical intrusion attacks. In [101] Kömmerling and Kuhn present some advice on how to design tamper resistant smart cards. In [102, 103] the authors present attack scenarios and attack classifications for smart card attacks. Information tends to leak over time and sadly this is also true for secret information. There are deep reasons for this, including information theoretical arguments and reasons related to the laws of physics. In practice the leakage reasons are more mundane, but leakages invariably occur. In [104] the authors deliberate on this issue and provide a practical context for the vulnerabilities.

The Mobile Termination Units
The mobile termination is responsible for handling mobile radio access. This involves running the radio access procedures and establishing a radio connection(s) with the network(s). It also involves invoking the mobility management procedures, including identity presentation and identity corroboration. The MT may provide access to multiple ANs simultaneously and these connections may be provided by different types of radio technology.

The MT must be able to protect the established radio links and the MT must therefore provide data confidentiality and data integrity functions. The MT will receive session key material from the UE.

The Clock and Positioning Device
The UD will certainly include a clock device (CLK). In Figure 4.4 the CLK device is depicted as an independent of the computing platform and of the

MT. The CLK device must be a protected device and setting of CLK time is a protected operation.

The UD may also have a positioning device (PD). The PD unit is assumed to be a protected unit, but the reported position will nevertheless be dependent on either the SN network or some other external infrastructure (for instance NAVSTAR/GPS or similar) to be able to determine the UD position. The PD should be able to indicate measurement resolution data and measurement dependencies upon request.

The Computing Platform
The Computing Platform as depicted in Figure 4.4 consists of user terminal functionality, external data interfaces and a secure boot and storage area. The "Secure Boot and Storage Area" should include functionality to allow remote device management.

4.5.3 The Serving Network and the Home Environment Entities

The SN provides a core network (CN) with one or more access networks (ANs) attached to it. The SN operates as one administrative entity and while a cellular network operator may also own HE entities, the SN shall be assumed to be a separate distinguishable entity. The SN, in this definition, does not have its own subscribers.

The HE is the entity that accepts user subscriptions. The HE does not, in this definition, itself provide access services. Rather, the HE negotiates access to SN networks on behalf of its subscribers through HE-SN roaming agreements. As has already been mentioned, cellular network operators commonly own and operate both HE and SN entities. One may therefore refer to a "Home SN" in order to distinguish this type of SN from an SN operated by an another operator (a "foreign" SN).

4.5.4 The Regulatory Actors

There will be different types of regulatory actors.

- **LEA** – "Law Enforcing Agency" (Lawful Interception)
  This actor will operate on the basis of the mandatory *Lawful Interception (LI)* requirements. Lawful interception is defined to be legally sanctioned access to private communications, such as telephone calls or e-mail messages. LI is a process in which a service provider or network

operator collects and provides the Law Enforcement Agency (LEA) with intercepted communications of private individuals or organizations.

- **PSAP** – "Public Safety Answering Point" (Emergency Call)
  The purpose of the Emergency Call requirements is to aid the user in times of distress. The ETSI Special Report on emergency call handling [105] is a useful reference for emergency call requirements.

- **DRD** – "Data Retention Directive" (Lawful surveillance)
  In contrast to the LI requirements, the data retention requirement applies to control plane data for all user traffic and to all user connections. The data is captured and stored without any *a priori* suspicion against the users. The motivation for the "Data Retention Directive" is prevention of crime- and terrorist acts [106].[3] The DRD is an EU specific directive, but similar schemes are in place (or about to come in place) in many non-EU jurisdictions.

The regulatory entities are not new to the cellular world. The LI and EC requirements have been in place for some time and they are already implemented for the 2G and 3G systems.

The EC requirements dictate that the calling line number, indication of the emergency caller's position be made available to the PSAP function (see [105, chapter 4.2]). The real-world LI requirement can be quite complex and the exact requirements are subject to national legislation. For the purpose of defining reference points in the communications model one can safely assume that both the HE and SN is required to provide LI interfaces. The "data retention directive (DRD)" represents a new type of regulatory imposed requirement. The effect is that the operators must be able to store a huge amount of call data for between 6 to 24 months. This begs the question of how one ensures data quality and how one protects the personal privacy of the subscribers. The question is important since the availability of these large DRD databases is problematic if the databases are compromised.

Additionally, for crime prevention purposes, the association between the IMSI (the subscription identity) and the IMEI (the mobile phone identity) are stored by the operators in some countries.[4]

---

[3] The final text of the EU Data Retention Directive (Council doc. 3677/05, 3.2.06) is available at http://www.statewatch.org/news/2006/feb/st03677-05.pdf.

[4] Storing the (IMSI, IMEI) tuple facilitates the tracking of stolen mobile phones and makes it easier to prevent mobile phone theft. The storage period is normally 3–6 months.

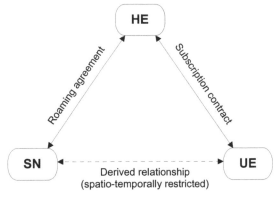

Figure 4.5 Trust relationships.

4.6 Trust Relationships

4.6.1 Trust Relationships and Control Requirements

Trust can be defined as an intersection between beliefs and dependence. The UE is clearly dependent on the HE to get service and the UE also has reason to believe that the HE is benevolent. Still, the UE should only trust the HE to the extent needed to get the required services. The relationship between the HE and the SN is normally well balanced, and while neither should have reason to distrust each other they should only trust their partner to the extent necessary for normal operation. The relationship between the UE and SN also requires trust between the parties, and here the HE will act as a mediator/proxy. Figure 4.5 depicts the trust relationships between HE, SN and UE.

Nominally, the trust relationships only involve the three given principal parties, but in reality the principals operate under constraints decided by society. The constraints are partly formed by legal and regulatory requirements, partly formed by public opinion and partly directed by considerations of the principals own (long-term) best interest (commercial interest etc).

4.6.2 The Long-Term HE-UE Relationship

The relationship is based on the subscription contract, which gives the HE security jurisdiction[5] over the UE. The relationship is therefore asymmetrical and the UE must necessarily trust the HE.

---

[5] The words *control* and *jurisdiction* may be used interchangeably in this context.

- *Security* – The HE assigns the UE identity ($UEID$) and the long-term security credentials. In this matter the UE is totally dependent on the HE.
- *Privacy* – The HE assigns the $UEID$. The UE therefore obviously cannot have *Identity Privacy* from the HE.
- *Control* – The HE is accountable for the UE while the UE is roaming. The HE therefore has a legitimate need for control over the UE with respect to incurred charges. The HE is also under regulatory obligation to maintain a certain degree of control over the UE.

4.6.3 The Long-Term HE-SN Relationship

The trust is restricted by the scope of the roaming agreement. The roaming partners are likely to be competitors in other markets, and they will not want to divulge more information than strictly necessary to the opposite party.

- *Security* – The trust is mutual. The trust should be limited to the actual needs.
- *Privacy* – The HE and SN operators will not want to disclose more information about their respective networks than strictly necessary. That is, the operators may want to conceal the location, address/identity and capacity of the various nodes in the network.
- *Control* – The HE operators will want to retain some control (Home Control) over the UE while it is roaming onto the SN network. The SN, which directly serves the UE, also has a legitimate need for control over the UE service usage etc. Charging and billing is a central issue, and the HE will want to limit the trust with respect to incurred charges.

4.6.4 The Temporary SN-UE Relationship

The SN-UE trust relationship is derived from the HE-UE and HE-SN relationships. While one may assume trust to be transitive, the derived relationship obviously has a weaker basis. Given the weaker basis it is advisable that the scope of the derived SN-UE relationship be limited. It is also reasonable to require online confirmation from the proxy (the HE) to assert the validity of the derived relationship.

- *Security* - Since the SN and UE do not have an *a priori* trust relationship, the derived SN-UE security credentials must be based on HE conformation/provision.

- *Privacy* - Whenever the SN is in radio contact with the UE, the SN will be able to derive the approximate UE position. The UE cannot therefore attain *Location Privacy* from the SN.
- *Control* - The SN-UE trust relationship depends on the HE and it is a temporary and derived relationship. It is therefore reasonable that both the SN and the UE requires online confirmation from the HE. It is also reasonable to define limitations on the SN-UE trust. This may be expressed as limited validity of the derived trust in terms of charging/billing and in terms of explicit expiry conditions.

4.7 The Security Contexts

4.7.1 Home Control

Figure 4.5 is slightly misleading in how the trust relationships work. The trust relationships between the principals are not fully bidirectional; In particular there is an asymmetry in the HE–UE relationship. The UE must unconditionally trust the HE; the HE cannot afford quite the same level of trust in the UE.

In the 2G/3G systems the HE delegates operational control to the SN. Technically this works fine, but it requires that HE have a high degree of trust in the SN. This may have been justified in the regulatory regimes and operator environments of the 1G systems and for early 2G systems, but the approach has become increasingly inadequate over time. In a future system the HE must therefore find a way to enforce a certain level of *Home Control* in order to protect the UE and its own interests.

4.7.2 Online 3-Way Authentication and Key Agreement Protocol

The 3GPP AKA protocols are two-staged off-line protocols [34, 39–41, 73] (Figure 2.7). Remarkably, the LTE (4G) security architecture, through backwards compatibility with UMTS, also depends on off-line authentication. A major problem with the 3GPP family of AKA protocols is directly related to the fact that the protocols are off-line protocols. The HE therefore cannot know when AKA events occur. This leaves the HE with very little control over the UE. One alternative to the basic 3G AKA was developed for the 3G-WLAN interworking case [107, 108]. The 3G-WLAN scheme is an adaption of the basic 3G AKA scheme. It provides online mutual entity authentication between the UE and the HE by running the *challenge-response* directly between the UE and the HE. The SN is then reduced to a passive party which just receives instructions and session key material. This is not always acceptable from the SN point of view. It is therefore asserted that the 2-party (UE – Network) schemes used in 2G/3G systems are inadequate. To rectify the situation one must design a solution where all three principals are active parties and where the control characteristics of the trust relationships are respected. To this end the AKA context must be established with all three principals online.

The three principals are not equals and the interactions between HE and UE is necessarily forwarded through the SN. We therefore do not have a straightforward 3-party case for a new AKA protocol. So, solutions that would have been appropriate for a 3-party case [109] do not automatically apply. Instead we have a special 3-way case. The one round-trip tripartite DH solution (for key agreement) developed by Joux in [109] (and refined in [110]) has many interesting qualities. However, the Joux solution assumes that "each participant is allowed to talk once and broadcast some data to the other two" (section 3 in [110]) is problematic in our setting since the (CC) channel passes through the SN. The SN not only *see* all messages over the (CC) channel, it also controls the (CC) channel. The Joux assumption about broadcast channels therefore does not hold for our case. As noted in [109, section 4] the exchanged parameters are relatively large and the computations will be relatively expensive in terms of processing overhead. The Joux tripartite DH solution therefore does not lend itself particularly well to the cellular topology and context. Finally, the tripartite protocols assume that the principals have public identities. This assumption is contradictory to our subscriber privacy requirement. Public-identity tripartite protocols are therefore deemed not to be viable for the cellular 3-way case.

4.7.3 Initiator-Responder Resilience

All key agreement protocols should ideally be Initiator-Responder resilient. That is, the premises for the authentication and key derivation should not be fully controllable by either party. The concept is explained in [111, section 2]. Our AKA protocol should aim at providing initiator-responder resilience. The resilience need not be 3-way, but should respect the trust assumptions outlined in Section 4.6.

4.7.4 Security Context Hierarchy

There is a natural hierarchy to the security contexts [8, 112, 113]. The hierarchy has both a temporal and a spatial dimension. Figure 4.6 depicts the

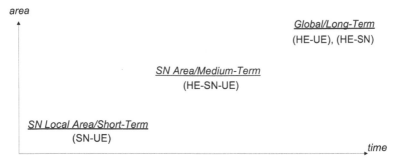

Figure 4.6 UE centric view: Security context hierarchy.

3-way security context hierarchy and Figure 4.7 illustrates the dependencies between the levels. It is noted that the security contexts depend on the system architecture and with different assumptions one may end up with different context hierarchies.

The Context Levels

Three context levels have been identified; the first being the long-term contexts, then there are the roaming contexts which represent an intermediate level and finally the session oriented contexts which are relatively short lived. Observe that one does not necessarily need all of the identified contexts.

1. *The Long-Term Contexts (LTC)*

   - $LTC_{(HE,SN)}$: The context is based on legally binding roaming agreements. It will normally apply to the full coverage of the SN network.
   - $LTC_{(HE,UE)}$: This context is defined by a subscription contract between the subscriber and the HE. The context contains long-term security credentials, including the permanent UE identity $(UEID)$. The context will exist for the duration of the subscription. The coverage area is in principle the combined area of all HE roaming agreement areas.

2. *Medium-Term Contexts (MTC)*

   - $MTC_{(HE,SN,UE)}$: This context is derived from $LTC_{(HE,UE)}$ and $LTC_{(HE,SN)}$ contexts. It is established dynamically by the 3-way AKA protocol in conjunction with the UE registration procedure. The validity of the context is determined by the SN area (server area or location area) and by an expiry period.

- $MTC_{(HE,SN)}$: This context is derived from the $LTC_{(HE,SN)}$ context and is independent of any specific UE. The validity is determined by usage (traffic volume) and an expiry period.
- $MTC_{(HE,UE)}$: This context is derived from the $LTC_{(HE,UE)}$ context, but may also be established by the 3-way AKA protocol.

3. *Short-Term Contexts (STC)*

- $STC_{(SN,UE)}$: This context is derived from $MTC_{(HE,SN,UE)}$ and it secures the access link (CA channel). It consists of symmetric-key key matrical for protection of user related communication and user related control signalling. The validity of the context is determined by usage (traffic volume and location).
- $STC_{(HE,SN)}$: This context is derived from $MTC_{(HE,SN)}$ and secures the CB channel. The context is established between security gateways. The HE and the SN may have multiple security gateways and there may be many simultaneously available CB channels.
- $STC_{(HE,UE)}$: This is an optional context between the UE and the HE (CC channel). It is assumed to be based on the $MTC_{(HE,SN)}$ context. It may be used for HE security management of the SM, UE and the UD computing platform.

Observe:

(a) The $STC_{(HE,SN)}$ context is seen as being independent of any particular UE. However, it is also possible to derive a UE specific context for the CB channel.

(b) The $STC_{(HE,UE)}$ context is seen as being independent of the SN, but $STC_{(HE,UE)}$ context can only be active when a $(UE \rightleftarrows HE)$ path exists. The $STC_{(HE,UE)}$ context may nevertheless exist independently of any specific $MTC_{(HE,SN,UE)}$ context.

Context Dependencies

As indicated in the context level description there is a chain of dependencies to the security contexts. Figure 4.7 illustrates the dependencies. As noted, other dependency chains are possible and may be suitable for different requirement sets.

Note that when a context expires, all dependent contexts should be recursively invalidated. The medium-term and short-term contexts should also have

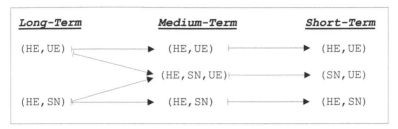

Figure 4.7 Security context dependencies.

additional explicit termination conditions. The principals involved should also be able to cancel the contexts prior to exceeding the expiry conditions. Upon an explicit cancel request the contexts should be invalidated immediately, and all communication associated with the security context should be stopped. If the HE terminates a $MTC_{(HE,SN,UE)}$ context then the HE must rely on the SN to act on the command and to forward it to the UE. The HE should log the termination command and it should not accept charging for events subsequent to the termination request. A similar case exists for UE initiated termination of the context.

Exposure Control and Spatio-Temporal Context Binding
Over time the security context will gradually be exposed and secrets may leak. We must therefore make sure that the security context expires before the exposure becomes critical. In traditional crypto-systems we have two exposure dimensions and correspondingly two types of expiry conditions: usage based and temporal. This is exemplified by IPsec [75], which can express *lifetime* in terms of both byte/packet count and time period. In a mobile setting one may add a third exposure dimension with spatially based expiry [6, 114].

(a) *Temporally Based Expiry*
   Cryptographic secrets/keys is susceptible to leak over time. It is therefore sensible to limit the lifetime.
(b) *Usage (traffic volume) Based Expiry*
   Cryptographic keys have limited entropy and one must limit the amount of data that is protected with a particular key.
(c) *Spatially Based Expiry*
   One may find reason to lock a security context to a geographically confined area.

The validity of the $MTC_{(HE,SN,UE)}$ context is naturally confined by a spatial validity (an Area Code (AC), related to registration area) and a validity period

(VP). The AC parameter is published on a broadcast channel. There may also be local area codes for the location/routing area (LAC), and this parameter can be used to further spatially qualify the $STC_{(SN,UE)}$ context.

The validity period, VP, should be long enough to avoid excessive context invalidation yet should not be substantially longer than normal UE presence in an area. The actual mobility patterns may vary over time and it should be possible to configure the VP parameter. The SN may publish its preferred validity period (VP_{SN}) on a broadcast channel. The UE may have a HE determined maximum (VP_{HE}) value, and the context should be canceled when one of the validity period requirements expires. That is, $VP = Min(VP_{HE}, VP_{SN})$. Synchronization can be a problem for the temporal expiry condition. The VP parameters should therefore be expressed in time units relative to the establishment event. The spatial expiry can only be controlled by the SN and the UE. The HE does not normally need to know the actual AC, and for location privacy reasons it is preferable if the AC/LAC is not revealed to the HE.

Enhanced Home Control

In some systems the HE may need strict home control. By setting the $MTC_{(HE,SN,UE)}$ expiry interval to be short (i.e. frequent AKA invocations) and by tailoring the usage expiry setting one can achieve tighter home control. The HE may also need to know exactly where the UE is located. The HE may then require that the UE/SN submit the location information to the HE. Note that the measurement rate may need to be adjusted according to UE velocity [114].

Instead of solving the problem of enhanced home control during authentication one may also implement a more generic access control mechanism, like the so-called Spatial Role-Based Access Control mechanism (SRBAC) [115]. Generally, a decoupling of spatial access control and security context establishment is advocated since it is not clear that spatial access control has sufficient benefits to warrant the overhead.[6] Furthermore, dedicated access control mechanisms, like the SRBAC proposal, can provide much better security policy resolution, fine grained access control and flexibility (including dynamic policy reconfigurations).

---

[6] The context establishment is highly delay sensitive in cellular systems.

Security Context Separation
In future mobile systems it is reasonable to expect the user device to connect to multiple access networks simultaneously. The different access types may have different levels of security and it is not advisable to use the same security context for different access types. One may conceivably have one common medium-term context ($MTC_{(HE,SN,UE)}$) that is used as a basis for derivation of separate short-term contexts ($STC_{(SN,UE)}$) for each access type. This is not without risk and there will be technical issues with context bindings and the spatial expiry condition. There is therefore a reasonably strong argument against sharing medium-term contexts between different access types, even if the argument is not conclusive for all cases.

A user device may also engage in communication with multiple SN operators. These SN operators may deploy the same access network types, but the security contexts are tied to the SN and separate security contexts will then be required.

Security Algorithms and Key Bindings
As demonstrated by Barkan, Biham and Keller in their attack on GSM [28], there can be severe consequences to not having a proper binding between cryptographic algorithms and session keys. The lesson to be learned is that keys must be bounded to the algorithm (dynamically or statically).

Recently it has also been demonstrated that the binding must additionally also be mode-of-operation specific. In [116] the authors demonstrate that using the same key for the same algorithm is not safe if different mode-of-operations are allowed to be used. Thus, the key binding requirement must be strengthened to require the key to be bounded to a specific (`algorithm`, `mode-of-operation`) instance.

The UMTS keys are susceptible to the binding problem, but in LTE one has proper binding including to the mode-of-operation (see Section 3.3.9).

4.8 Performance Consideration

4.8.1 Security Context Establishment

The medium-term context ($MTC_{(HE,SN,UE)}$) is established by a 3-way AKA protocol. The participating principals face different constraints, both computationally and with respect to the communication channels. The AKA protocol is logically part of the location registration procedure and is executed in response to external events and user interactions; in both cases timely procedure completion is essential for the perceived performance of the system. The AKA protocol must take this into account and make sure that the AKA is well balanced to avoid computational and communications bottlenecks in the protocol execution.

4.8.2 Computational Balance

The Secure Module

The UE must be able to compute the AKA functions on-the-fly. Modern mobile devices are capable of executing all the required cryptographic functions with ease. This may even be true if we assume use of computationally expensive algorithms including Identity-Based Encryption (IBE) transformations, Diffie–Hellman (DH) exchanges or Secure Multi-party Computation (SMC) methods.

Still, one must ensure that the AKA protocol is balanced such that the SM is not overloaded.

The Serving Network

The SN must participate actively in the future AKA protocol. The SN nodes will serve a comparatively large number of users and must be able to execute the AKA protocol on-demand. The instantaneous processing requirements may therefore be demanding. Pre-computation may be possible for some operations (for instance by pre-computing DH parameters). Use of dedicated hardware based crypto-accelerators may be required.

The Home Environment

The HE nodes will serve a large population of subscribers. To instantly compute security context credentials may create a substantial load. The HE nodes must therefore be dimensioned for a high instantaneous crypto-processing load.

The HE may be able to spread the load to lessen the impact of processing peaks. This could be achieved by having an execution window for context renewal events. The HE may also exploit the possibility for pre-computation, which would be possible for computing DH parameters, etc.

4.8.3 Communication Balance

The A-Interface (CA Channel)

For the control plane our concern is focused on the round-trip delays. The propagation delay is a concern; and in particular the latency in the channel setup can be problematic. The radio channel is a shared physically restricted resource and there will necessarily be capacity limitations. Control plane signalling is not very demanding capacity-wise and under most circumstances there should be no problems here. However, there may be restrictions on the Message Transfer Unit (MTU) size during access signalling. For performance reasons one must try to avoid message segmentation and the MTU size may be a limiting factor in the AKA protocol design. The MTU size restriction may preclude or complicate support for crypto primitives that requires large information elements, e.g. for Diffie–Hellman based key generation and asymmetric crypto systems with large block size. Support for data expanding primitives ($E_K(M) \rightarrow C$, where $|C| \gg |M|$) may likewise be problematic. There may therefore be some design restrictions on the AKA protocol over the A-interface.

The B-Interface (CB Channel)
The B-interface is expected to be a fixed network interface. There should be no capacity problems for the B-interface. One should still strive to contain the communication delay over the B-interface. This problem can be solved by having dedicated roaming networks with a reasonably low number of intermediate servers/routers. This type of roaming networks already exists for the 2G/3G cellular systems and the GSM Association has released a guidelines document on 2G/3G roaming networks [117].

4.8.4 Performance and Integration Issues
Mobility Management and Security Protocol Integration
The security procedures must be designed to meet the overall system performance requirements. The accumulated delays during initial registration and user session establishment are of particular importance. An inspection of the 3G system call set-up signalling scheme reveals that the mobility procedures are mostly evolved 2G procedures. The 2G procedures were not at all bad for the 2G systems, but the 2G service model was focused on providing telephony type connections, i.e. the system was focused on providing circuit-switched connection-oriented services. In addition, there were many technical limitations due to the age of the 2G system architecture. The control model for future mobile systems should therefore be designed from scratch, and the main input should be the system service model.

Integration is possible in many areas. The Mobility Management *MM Location Registration* procedure will naturally coincide with invocation of the AKA protocol. It is also observed that the *MM Identity Presentation* procedure must precede execution of the AKA protocol and that it would be beneficial to consider the procedures together. It is argued that the total number of round-trips can be reduced, even when deploying an online privacy enhanced 3-way AKA protocol. There is an important consequence of integrating the *MM Location Registration*, *MM Identity Presentation* and the AKA protocol; namely that the sequence must necessarily be initiated by the UE. This is because it is the task of the UE to initiate *MM Location Registration*. In the 2G/3G systems the network is always the initiator of the AKA protocol, but it is advised that the new AKA scheme is changed to allow tighter integration and thereby enhanced performance.

There is also integration potential for the establishment and renewal of the short-term context ($STC_{(SN,UE)}$). For instance, channel re-allocation may trigger re-keying ($STC_{(SN,UE)}$ negotiation). The $STC_{(SN,UE)}$ negotiation can perhaps best be provided as a functional command where the information element block would be transferred as a parameter to a channel re-allocation command.

4.9 Link Layer Protection

The key material to be used for link layer protection is derived from the AKA security context(s). The link layer protection termination points are important since any divergence from the AKA termination point means that there must be schemes for protected key material transfer in place.

4.9.1 Directional Keys

In the 2G and 3G cellular systems the keys used are bidirectional, i.e. the same keys are used for protecting both uplink and downlink communication. This was a reasonable strategy for circuit-switched connections in which the transfer is normally symmetrical. For a packet-switched service model one may want to revise this strategy. There is in fact a strong case for having directional short-term security contexts (STC) between the UE and the SN.

4.9.2 Control Plane vs. User Plane

A point worth noting is that the security termination points of the user plane and the control plane need not coincide. It may therefore be beneficial to provide separate security contexts for control plane and user plane data. One may also decide to differentiate radio access protection from network access protection. For instance, in the 3GPP LTE/EPS architecture one separates control plane and user plane, and one provides independent AN and CN protection (see Section 3.3).

4.9.3 Confidentiality Requirements

The data confidentiality protection should, for efficiency reasons, be based on symmetric crypto-primitives. When it comes to issues like block length and key length it would seem wise to comply with recommendations from (authoritative) sources like the "ECRYPT Yearly Report on Algorithms and Keysizes (2007–2008)" [118].[7]

The ECRYPT report takes into account aspects such as Moore's law and time-memory-data trade-off etc. In table 7.1 of the ECRYPT report [118] the recommendation is for a minimum of 84-bit key size. This would assume a very powerful attacker and the protection should be effective for at least a couple of months. One must take the lifetime of the system into account when deciding on key length. Typically, the estimate for a cellular system is for a lifetime of around 20 years.[8] If one assumes that Moore's "law" of a doubling of processing power every 18 months holds for the next decades then one needs approximately 13 additional bits for a 20-year lifespan and approximately 20 additional bits for a 30-year lifespan. This would add up to a key size of approximately 100 bits. This would suggest that the current industry standard of using 128-bit keys is a reasonable choice.

To increase the number of bits in the key is not necessarily effective as demonstrated in the case of the Wired Equivalent Privacy (WEP) [119]. To safeguard against crypto-primitive failure one must therefore design the system to permit algorithm deprecation. This means that the security architecture must include secure algorithm negotiation.

4.9.4 Integrity Requirements

The actual requirements will depend on the access link to be protected. The integrity protection should be designed in conjunction with the data confiden-

---

[7] ECRYPT – European Network of Excellence for Cryptology is a "network of excellence" program within the EU IST programme. More information at http://www.ecrypt.eu.org/.

[8] According to GSM Association (http://www.gsmworld.com/about/history.shtml) the basic GSM system parameters were agreed in 1987. In this sense GSM is already more than 20 years old.

tiality protection. The data integrity protection should, for efficiency reasons, be based on symmetric crypto-primitives.

When it comes to issues like block length and key length it would seem wise to comply with recommendations from (authoritative) sources like the ECRYPT recommendations [118]. As would be expected, the ECRYPT report distinguishes between the key size requirements and the checksum size. There is no specific recommendation on key size and the arguments for minimum checksum is similar to those found in Section 2.6.4. In the ECRYPT report there is a general recommendation to use the same key size as for the confidentiality primitive. Thus, a key size of 128 bits is a safe choice for the data integrity primitive.

The integrity protection requirements may be different for control plane protection and user plane protection. For control plane protection the environment scope is to provide protection of real-time signalling data. As was concluded in Chapter 2 the current 32-bit checksum used in 3G systems is dangerously short and it may be wise to increase the length to 64 bits given the lifetime of the system.

For the user plane the message lengths tend to be longer. This may justify a longer checksum. The validity of the user plane data is likely to be significantly longer than for the control plane. Furthermore, increasing the checksum is not free and the incurred communication overhead must be justified. High-value transactions may need enhanced integrity protection and would probably require end-to-end protection. This is outside the scope of the link layer protection.

4.9.5 Combined Methods

It may be permissible to use combined methods in which confidentiality protection and integrity/message authentication protection is achieved by one mode of operation.

An example of combined methods is found in NIST Special Publication 800-38D [120]. NIST SP 800-38D specifies use of Galois/Counter Mode (GCM) with the AES algorithm. GCM combines the counter mode for confidentiality with message authentication that is based on a universal hash function. GCM is primarily intended for high-throughput applications which needs relatively large checksum tags (between 96 and 128 bits).

4.9.6 Protection Range

Security Termination Point and Key Distribution

In a real network there will be many interfaces and a large number of physical nodes. Our model with two interfaces and three principals (Figure 4.2) is too limited when it comes to discussing the issue of security termination point. In a future cellular system the access points will be highly distributed and they will generally not enjoy the safety of a protected location. So, from a security point of view the AP may not be a good security termination point. However, in LTE/EPS one does terminate security in the eNodeB, due to stringent radio related requirements.

In GPRS the termination point was moved to the core network (the SGSN). The problem with that is that from a system design perspective the core network should *not* need to know about the specifics of the ANs. When security is terminated at the core network this principle is violated. There are also other problems with quality measurements and AN monitoring that will then have to be routed via the core network, which is problematic from a radio performance perspective. So access security should preferably terminate somewhere in the AN.

4.10 Know Thy Enemy

There are many possible attack vectors for an intruder in a future cellular system. For the communication channels the traditional worst-case scenario is for the so-called Dolev–Yao intruder [100]. The DY intruder can selectively read, store, delete, inject and otherwise manipulate all transmitted messages (in real-time). The DY intruder may also be a legitimate principal and it may try to impersonate any of the other legitimate principals. The DY intruder model is not the only conceivable intruder model and it does in fact not represent the most powerful intruder either. One may easily envisage intruders that are able to physically compromise the entities or that may by other means gain access to confidential information belonging to the entities.

4.10.1 Threats, Risks and Attack

Threats and Risks

The threats and risks facing a future mobile system is assumed to be a superset of the threats and risks that faces the current 3G systems. That is, the high-level threats defined in 3GPP TS 21.133 "Security Threats and Requirements" [38] is basically assumed to be valid also for a beyond-3G system.

For a real system there will be refinements, adaptations and amendments to the contents of TS 21.133, but for now the threats defined in TS 21.133 are provisionally accepted.

The risks facing a beyond-3G system is to some extent different from the risks facing a 3G system. One should expect the future system environment to be even more hostile. Technological progress will make it easier to execute (what is today) expensive attacks. For example, for the 2G systems it was initially deemed impractical to launch false-BTS attacks. Today, this is within the scope of any dedicated intruder[9]. Another example is the attacks against the GSM A5/1 algorithm (see [27, 28]). These attacks would require a resourceful and dedicated intruder when the attacks were first published, but cheaper and more capable computing platforms means that this equipment would now be so cheap as to be affordable to a large number of people.

The lesson to be learned is that if there is an attack that may succeed if more resources is assigned to it then one should expect the attack to be feasible and maybe even practical within the lifespan of the system. There is obviously still some attacks that are so hard to mount that (brute-force) technological progress alone will not matter.

Attack Types

The generic intruder must be assumed to use all possible opportunities to execute attacks. The Dolev–Yao intruder is limited to attacks on communications channels, but it will be able to access all communications channels. The following attack types are then possible:[10]

- *Eavesdropping*
  The adversary captures information from the channel. This is a `read` type of attack.
- *Insertion*
  The adversary insert information to the channel. This is a `write` type of attack.
- *Modification*
  The adversary alters information sent over the channel. This tends to be a `read-then-write` type of attack, but may also be a `write` attack (direct overwriting).

---

[9] It requires acquiring and controlling a BTS and so it is not an amateur hack.

[10] The lists are derived from table 1.3, section 1.6 in "Protocols for Authentication and Key Establishment" by Boyd and Mathuria [121], but several additions and amendments have been made.

- *Deletion*

  The adversary deletes information sent over the channel. This can be considered a `write` type of attack (overwrite to make the information unintelligible).

It is noted that in general it is difficult to execute *modify* or *delete* attacks over a radio channel. It is not that the intruder cannot attempt these attacks, it is simply that in a radio environment is difficult to ensure that the attack is effective. In particular, the *modify* attacks are difficult to stage. Jamming the channel with noise, *delete* attacks, are certainly possible, but they would likely only achieve disruption of the communication channel (Denial-of-Service). Due to the physically distributed environment, all attacks must additionally be considered to have a spatially confined scope.

The following list consists of more sophisticated attacks; attacks which may be seen as compound attacks constructed from the primitive attacks:

- *Replay*

  The adversary records information from the protocol run (or a previous protocol run) and then retransmits the data (to one of the principals). This is a `read-wait-write` type of attack.

- *Preplay*

  The adversary engages in a protocol run with one (or more) of the principals such that the adversary participates in a protocol run prior to some other legitimate principal. This is an attack where the aim may not be to break the protocol at the first try, but rather to gather information (which presumably is utilized in a subsequent attack). This a compound attack with multiple `read` and `write` operations.

- *Reflection*

  This attack involves the adversary reflecting messages back onto the sending principal. It is a special case of the *replay* attack.

- *Man-in-the-Middle (MitM)*

  The MitM attack is an attack in which the adversary is located between the legitimate principals. Typically the MitM adversary is capable of intercepting messages and then selectively modify/delete/insert/replay/preplay/forward messages at will. If successful, the MitM attack typically fools the legitimate principals into believing they have secure direct communication between them whilst in reality the adversary is controlling all communication.

- *Denial-of-Service (DoS)*

  By itself a DoS attack will only disrupt the communication between legitimate principals. This may be an end in itself, but DoS attacks may also be part of a more sophisticated attack. For instance, if the adversary launches a covert DoS attack that prevents normal communication over a secure channel, then the involved principals may be fooled into setting up a less secure channel and conduct their transactions over that channel. Observe that this type of attack may be particularly successful if the legitimate principals are tricked into believing that it is the security configuration that makes the transaction fail.[11]

There are several types of DoS attacks, including Distributed DoS (DDoS). Physical DoS attacks on the A-interface will necessarily be spatially confined. To completely block a RAN will therefore be difficult. Logical DoS attacks may be more feasible, but it is probably still the case that DoS attacks which target the lower layers will not scale well on mobile systems. Nevertheless, selective DoS attacks (per cell or per access attempt) may be easy to execute and may be highly effective. It is also noted that many DoS attacks consist of flooding the target with perfectly legal and valid requests. The protection strategy must take this into account. Furthermore, some DoS attacks exploit the asymmetry in principal capability and workload during a protocol exchange. Security protocols may be vulnerable to this type of attack since they commonly require execution of computationally demanding cryptographic operations.

- *Typing attacks*
  The adversary successfully replaces a message element with another message element of different type. Normally both message elements are encrypted, and so this attack may mislead the principals into using the wrong plain-text. The efficiency of this type of attack would be dependent on how the cryptography is being used.This attack type will generally be foiled by proper use of data integrity protection. This is a `read-then-write` type of attack, and is an advanced special case of the replay attack.

- *Crypto-analysis*
  Crypto-analysis is an attack type in which the adversary is able to break the cryptography. The DY intruder cannot break crypto-primitives, but it may still attack improper use of crypto-primitives.

---

[11] For instance, if an attack succeeds into making the principal believe that the firewall is blocking too much then the principal may be tempted to turn off the firewall. Likewise, if the adversary can make IKE security-association negotiations fail then the principal may be tempted to use old keys or turn off IPsec altogether.

For a more extensive treatment of attack types the reader is referred to the Boyd Mathuria book [121]. Alternative flaw classification schemes are found in the papers "Cryptograph Protocol Flaws" [122] and in "Cryptographic Protocols over Open Distributed Systems: A Taxonomy of Flaws and related Protocol Analysis Tools" [123].

Denial-of-Service Attacks against the Security Protocols

Some DoS attacks are targeted specifically against the security protocols. The AKA protocol must operate in an environment where it is very easy for an adversary to carry out access-denial DoS attacks simply by disrupting the radio transmission. These attacks cannot be prevented by the AKA protocol.

An adversary may try to block the channels by flooding the channels with "invalid" request as was the case with the TCP-syn attack [124]. However, this is significantly more complex than simply disrupting the channel by radio noise and the effect would likely be similar. Many of the access-denial attacks are local in nature, and in contrast to the TCP-syn attack they do not scale well. It is therefore questionable if we should make an effort in trying to avert this type of attack at the AKA protocol layer. The exception being that one must make sure that logical attacks cannot block specific users from accessing the system. An intruder may for instance impersonate a mobile subscriber and wilfully fail the authentication on repeated attempts. Security protocols commonly restrict the number of unsuccessful authentication attempts. After having experienced N unsuccessful authentication attempts the user may be blocked from making further attempts. To prevent this type of blocking attack from being successful it is essential that the blocking (if any) be restricted to a limited period.

One way to limit the effect of computational DoS attacks is to let the SN restrict the arrival rate of AKA invocation per access point. The HE, likewise, may limit the number of simultaneous AKA sessions from any given SN. Together, this will prevent a computational DoS attack from scaling. Needless to say, the SN and HE must monitor the number of unsuccessful access attempts due to authentication failure, etc., as it may be a sign that the system is under attack.

4.10.2 The Enemy Within

Dishonest Principals

The intruder models normally also take into account misbehaving principal entities. That is, a legitimate principal may try to impersonate as another principal or otherwise cheat on its communications partners.

The communications model (Figures 4.2 and 4.3) makes it inordinately difficult for the UE to fool the SN into believing the UE is a HE. Similarly, the SN will not likely be fooled by the HE into believing that the HE is a UE. It is therefore asserted that the SN will be able to determine whether a message originated from the A-interface or the B-interface. Furthermore, it really does not make sense to assume that the HE should impersonate the UE. The HE does after all have security jurisdiction over the UE, and while this does not automatically make the HE omnipotent with respect to the UE, it should be clear that the SN would have to trust the HE to be honest with respect to the UE. This is captured in the trust model assumptions (Section 4.6).

Multiple Dishonest Principals

Conceivably, there also is the case that two of the principals cooperate cheat on the third principal.

A) (UE, SN) cheats on HE
   Here one may assume that there are multiple UEs (belonging to the HE) which team up with the SN. The goal may conceivably be to impersonate other UEs.
B) (UE, HE) cheats on SN
   From the SN point of view the UE are merely an extension of the HE. The goal would presumably be to gain free access to SN services.
C) (HE, SN) cheats on UE
   In this case there would be very little the UE can do about it.

All the above cases may be realistic under some circumstances, but both A) and C) require the UE[12] and HE to cheat on each other. They will not normally do this unless the principals have been compromised (or partially compromised) in some way.

---

[12] It is important here to strictly distinguish the user/subscriber from the UE. The user/subscriber could well want to cheat, but the UE is normally assumed to be honest.

4.10.3 Effective Cryptographic Primitives

The intruder models tend to be qualitative and not quantitative. The models therefore often ignore issues with key length and key usage (lifetime issues). If the primitives are broken then the security of the system is trivially broken. From a modeling perspective one tends to assume that the cryptographic primitives are either effective or totally broken. Should the cryptographic primitives be broken, from a modeling perspective, one must consider data to be unprotected.

However, in practice the offered protection may suffice even when there are flaws in the cryptography. In [28] the authors demonstrated how to break the GSM crypto-system and the GSM crypto primitives. Despite this, the GSM security scheme is still successfully protecting millions of calls every day. The attack is effective, even if it does not scale particularly well on a system level. From a qualitative (modeling) perspective the GSM crypto-system is nevertheless broken. Sophisticated attacks on the crypto-mechanisms, like advanced attacks on the mode-of-operation of a cipher, it cannot be modeled with the traditional DY intruder models. Naive or incorrect use of crypto-mechanisms may be still be modeled with the standard intruder models.

4.10.4 The Standard Dolev–Yao Intruder Model

The standard Dolev–Yao intruder model [100] is a simple but powerful model. The DY intruder is capable of intercepting all communication channels and the DY intruder can manipulate all messages at will. The generic DY intruder also has the capability of storing all data ever exchanged on the communication channels. The DY intruder will miraculously be able to use its knowledge and abilities such that if an attack is theoretically possible then the DY intruder will execute it.

The DY intruder will attack and intercept on live sessions, but it may also use recent information to go back in time and retrospectively break old protocol runs. This may allow the intruder to learn even more and thus improve future attacks. It may obviously also allow the DY intruder to break the confidentiality of old message exchanges and it may allow the DY intruder to falsify these historic protocol runs.

The DY intruder is clearly a very powerful adversary. Still, there are limits to the DY intruder powers. By definition a DY intruder is not capable of physical intrusion of the principals. Another point worth stressing is that while the DY intruder can manipulate messages at will, it cannot actually break the cryptographic primitives. Note that while the DY intruder cannot break the cryptographic primitives *per se*, it may still be able to break the crypto-system. In fact, there is no shortage of examples of this type of crypto-system failures. The paper "A Survey of Authentication Protocol Literature: Version 1.0" by Clark and Jacob [125] presents a large body of examples of failed authentication protocols, where crypto-system failure is a major failure cause.

4.10.5 Subsets of the Dolev–Yao Intruder Model

The Wireless Intruder

In the standard DY model the intruder is capable of intercepting all channels. In practice, all channels are not equally exposed and it may also be the case that some of the principals have *a priori* successfully negotiated a secure channel between them.

In the real-world there are also physical limits to what an intruder can realistically achieve. For instance, in a wireless environment it will generally be hard to selectively delete information and it may be even harder to selectively replace information on a geographically distributed channel. It may therefore make sense to try to reduce the capabilities of an intruder in a wireless environment. The wireless-only intruder may be able to delete information, but will likely not be able to always do so.

In a highly synchronized real-time environment, which the radio environment represents, it may additionally be physically impossible to selectively modify information. Even selective deletions may be impossible to achieve. Another issue is whether the information can be deleted/modified without the legitimate principals noticing. In a radio environment it may be difficult to control the noise/signal propagation, and the principals may thus be alerted to the anomalous state-of-affairs. An attack that is easily noticed by the principals is less likely to be effective, and thus the attack may achieve little but disrupting the communication.

Another aspect of a wireless-only intruder is that it will likely be limited in its capability to read all channels. That is, due to the distributed nature of the wireless network the wireless intruder may only be able to read the channels in its geographical vicinity. In a broadband system the intruder would have to log an enormous amount of data. All this information would have to processed by the intruder and taken into account for every possible attack vector, potentially in real-time. This is obviously a very difficult trick to pull off. It is therefore realistic to assume that a real-world intruder can only use information from a limited period of time in real-time attacks.

Capabilities of the Wireless Intruder

The following is a short-list of restrictions a wireless-only intruder may face:

- *Pre-secured channels*
  The intruder may not be able to access all channels or even know that certain channels exists
- *Limited delete capability*
  The intruder may be hindered by the physical environment and radio system technology in deleting selected information elements. Or alternatively, the intruder cannot delete information without alerting the legitimate principals.
- *No selective replacement capability (i.e. no modify capability)*
  The intruder may be hindered by the physical environment and radio system technology in modifying selected information elements. Or alternatively, the intruder cannot modify information without alerting the legitimate principals.
- *Spatial restrictions*
  The intruder may only be able to operate within a confined geographical area.
- *Temporal restrictions*
  The intruder may only be able to store and apply a short history of recorded protocol runs. The restriction is likely to affect the capability for real-time attacks, but storage/processing limitation may also prevent the intruder from effective use of historic data.

This discussion is by no means meant to be exhaustive, but should serve to illustrate that subsets of the standard DY intruder are conceivable and in many ways more realistic than the full DY intruder.

The ideas for the wireless-only intruder presented above are not unique. For instance, in the AVISPA EU IST project one has envisioned an "over-the-air" (section 3.4 in [68]) intruder with comparable characteristics to the wireless-only intruder.

4.10.6 Supersets of the Dolev–Yao Intruder Model

Attacks against the Principals

There are many possible extensions to the standard DY intruder model. One extension would involve the capability of the intruder being able to compromise honest principals. This ability would primarily be due to physical intrusion attacks, but the means is not necessarily important. Computer hacking, social engineering, bribery of employees, etc., are other feasible means.

The capability to compromise principals may not be universal, and the intruder may have to choose its victims opportunistically. The principals are not equally vulnerable to intrusion attacks. For instance, the UD/UE may be far easier to penetrate and compromise than the HE. The UDs are highly distributed and potentially exposed to adverse environments. The physical integrity of the UD will therefore be hard to ensure and it may not be economically viable to deploy better physical protection. In the paper "Mobile Terminal Security" [102] the authors describe the issues and the trade-offs involved. See also Section 4.5.2 for a discussion of the security module and the associated secure module (SM).

The HE nodes will likely be well protected with professional operations and management procedures in place. Still, it is noted that the potential damage suffered by HE compromise is far greater than a single UD/UE compromise. The SN is also likely to be well protected, but the access network will have nodes (cells/access points) that must necessarily be geographically distributed. The RAN nodes are therefore likely to be more exposed than the core network nodes.

Cryptography Breaking Capabilities

The standard DY intruder does not break cryptographic primitives.[13] It is a bit ironic since the capabilities of the DYI will otherwise scale perfectly, even with large amounts of data (from previous protocol runs etc). So it would seem that the DYI is otherwise capable of handling combinatorial complexity. An enhanced intruder may have some crypto-breaking capabilities. For instance, even if the crypto-primitive is effective it may be vulnerable if the key size is too small.[14] Crypto-primitive breaking attacks may not scale very well, but they could be very effective against selected targets.

---

[13] All cryptographic primitives can be broken by (exhaustively) searching the key space. Other more effective attacks may also exist, for instance attacks that exploit weaknesses in the algorithmic construction of the crypto-primitive.

[14] One well-known crypto-primitive breaking effort is the breaking of RC5-64 algorithm arranged by *distributed.net*. The story is documented on http:/www.distributed.net. Another example is the DES cracker designed by the Electronic Frontier Foundation (EFF) [126]. The DES cracker is also documented at http://www.eff.org/Privacy/Crypto/Crypto_misc/DESCracker/.

Personal Privacy and Information Leakage

Some types of information leakage will be very hard to defend against [104]. For instance, information relating to the presence of a principal in a certain area at a certain time, the fact that the principal is engaged in some kind of transaction, etc. There is a physical underpinning and physical representation of the information, and in some sense one cannot truly prevent an intruder from learning some of the facts about the principals and the communication between them.

In practice this may be most prominent for subscriber privacy aspects. In the standard DY intruder model nothing is generally made of the fact that the DY intruder may learn the identities of the principals. For entities that must necessarily be visible, like the public HE and SN networks, this does not matter much. However, for the subscriber (the UE) it may well matter that information about the subscriber identity and subscriber location is leaked to an intruder. Even worse, a powerful intruder may well be able to trace the subscriber for a prolonged time and is thus able to determine the mobility pattern of the subscriber. This may in turn be used for association attacks (affecting the privacy of the UE). Identity theft can of course also be facilitated with the failure to adequately protect the UE identity.

To some extent one cannot prevent an intruder from tracing the radio signal information. At the lower layers the intruder may use all available knowledge of the transmitted signal and it must be assumed that an advanced intruder is able to determine the location of the subscriber. If the surveillance is sufficiently complete, by the aid of sensor networks etc, then it would be almost impossible for the subscriber to avoid UD(MT) tracking. The subscriber may use randomized and frequently switched anonymous identities, but the physical information leakage may betray the subscriber anyway.

The extended intruder may not know the true identity of the principals. At another level, the fact that there is a communication protocol means that there will be regularity to the exchanged information. Over time a pattern will likely emerge and the pattern may be distinctive enough to extract a unique reference. Such an acquired reference would have similar qualities to that of an identity, although this type of referential identity may only be meaningful to the intruder. The derived reference may enable the intruder to track even anonymous users for prolonged periods of time. Statistical information leakage may also include information about the service types consumed. It may also be possible to determine aspects like web site visits by using advanced traffic flow analysis and by using data mining techniques. Some of these aspects can be determined even if the data is confidentiality protected.

4.10.7 Choice of Intruder Model

The Wireless Intruder vs. The Standard DY Intruder

In a wireless setting it does make sense to limit the power of the intruder. It will be *extraordinarily difficult* to completely control the physical environment to such degree that the intruder can, at will, modify or delete the data transmitted/received over-the-air. So to assume an intruder that can corrupt and insert data, but which cannot delete/modify transmitted data seems sensible. Assessing the capabilities of a wireless-only intruder is difficult. Any reduction in strength must be thoroughly assessed and documented. The reduced capability set must be judged very carefully to avoid invalid assumption. Even if one insists that the wireless-only intruder has a reduced capability set one must be cautious to assume that this truly translates into reduced attack capabilities. The wireless-only intruder may appear to have limited capabilities, but as reported in the paper "Dolev–Yao is no better than Machiavelli" [127] such appearances can be deceptive. It is therefore obviously safer to assume a full DY intruder. Qualitatively, it is also clear that a system that withstands the full DY intruder will withstand the wireless-only intruder.

The DY Superset Intruder vs. The Standard DY Intruder

A DY superset intruder can easily be defined. However, it may not be so easy to model such an intruder. It is perhaps best if the extended DY intruder capabilities be modeled statistically. That is, the model would be more quantitative than qualitative. Either in the sense that the intruder is limited to the number of targets or that the non-DY attacks have success probabilities. For instance, there could be an $N\%$ chance of breaking a 64-bit cipher within a certain time-frame given the use of a certain amount of resources. The defense mechanisms should be based on risk analysis assessment and cost-benefit considerations, and the defense would to some extent focus more on impact reduction and mitigation of an attack rather than prevention.

The ability for the intruder to gather all information on all layers of the protocol stack and to take into account spatio-temporal information is easier to model quantitatively. However, the intruder models are usually given at an abstraction level where the amount of available information has been reduced to the bare minimum as the complexity of model state-space tend to increase rapidly (combinatorially) with the number of information elements in the model. So while it may be possible to define a spatio-temporally extended DY intruder, it may not be a practical model with respect to automated evaluation.

Table 4.1 Attacker classification.

Intruder	Resources	Comment
Hacker	Desktop PC; $400 budget	Single person intruder
Small organization	Multiple PCs; $10.000 budget	Experts available
Medium organization	Many computers; $300K	Many experts available
Large organization	Many computers; $10M budget	Many experts available
Intelligence Agency	Many computers; $300M budget	Many experts available

4.10.8 Realistic Intruder Models

Available Resources

The Dolev–Yao intruder and variants thereof is not the typical adversary. A classification of realistic would-be intruders and their respective budget is presented in [128]. The classification is assumed to be largely valid today [118, section 4.4], but one may assume the potential number of intruders in each class has increased considerably. The classification is amended slightly from the original in table 1 in [128], which also included cost estimates in time and money for breaking 40 and 56-bit keys (valid as of late 1995). Today this information is of historic value only.

Table 4.1 refers to a budget in dollars. In the original table this was described as a budget to be used on FPGA and/or ASIC hardware. Today one may also opt to buy hardware with dedicated crypto-primitive support (crypto accelerators etc). More recently we have also witnessed the move towards multi-core CPU design and the use of very powerful graphics processing units. These designs are very well suited for crypto-breaking efforts. For an attacker to succeed in a wireless setting it will also be necessary to spend money on radio equipment in order to intercept the radio traffic.

Who Is Being Attacked

It is worthwhile noting that some attacks will primarily be an attack against the subscriber while others can be considered to affect the HE and/or the SN the most. The operators, HE and/or SN, will have comparatively large resources to draw on to defend themselves against attacks. The UE can do little itself and is dependent on the HE and to some extent the SN to provide sufficient protection.

For attacks that primarily affect the UE and where there is little impact on the networks as such one should not assume that the network operators will spend too much time or effort on defending the UE. This is not to say that the network operators do not care about the subscribers, but one should be realistic here. The operators will be compelled to provide a reasonable level of protection for their subscribers, but the focus will be on cost effective protection for the population of subscribers rather than on protecting individual subscribers.

4.11 Summary

In this chapter a number of aspects that influence the design of a future mobile system are investigated. Mobility management is a very important topic for mobile systems. The mobility management protocols deal with setting up and managing the subscriber context at the serving network. The *subscriber registration* part of the mobility management shares many features with the security setup and the procedures are largely triggered by the same physical events. A section was devoted to analyzing a system model for a future mobile system. The model is not the only possible one, but it does have many of the features that one expects from a future mobile system. Complementary to the system model is the simplified communications model presented in Section 4.4. The principal entities and regulatory entities were also discussed. The nature of the regulatory actors may differ in the various regulatory and jurisdictional regimes, but the true purpose of the description is to illustrate that public cellular networks will have to accommodate regulatory requirements like support for emergency calls, lawful interception, etc.

An important issue in the design of a security architectures is the trust relationships. Security contexts were also discussed and a security context hierarchy was developed. Based on this a set of security contexts was outlined.

Performance is always important and the topic is discussed in Section 4.8. Several aspects must be taken into account, including computational aspects, communication capacity constraints and the number of total number of round-trips in the registration process. As it turns out, there is a considerable potential for reduced accumulated round-trip delays if the location registration and security context establishment is combined. The combined scheme would include UE identity presentation, authentication and key agreement and transfer of the user service context (subscriber information). The performance improvements can be realized only if the initiator-responder model found in the 2G and 3G systems is reversed.

Various aspects of link layer protection were discussed and analyzed. This includes issues such as security service needs, security services strength (key sizes, etc.), security termination point in the network and various integration issues.

Finally, the section "Know Thy Enemy" was dedicated to the threats, risks and possible attacks a mobile system may suffer. This discussion started off with the common Dolev–Yao intruder model, before investigating other possible models including both subsets and supersets of the DY intruder model. The section concluded with a brief look at the classification of the capabilities of some realistic adversaries and a perspective on the fact that the objectives of the offered protection depend on the vantage point.

5

Privacy Matters

> I have as much privacy as a goldfish in a bowl.
>
> *– Princess Margaret*

5.1 Introduction

The current 3G cellular systems have, despite some shortcomings, achieved a reasonable level of subscriber data confidentiality and data integrity protection over the wireless link. However, when it comes to provision of identity privacy and location privacy the 3G systems have not fared well. Provision of a credible subscriber privacy scheme is therefore an obvious area of improvement with respect to the current 3G systems. The LTE subscriber addressing model is identical to the 3G/UMTS addressing model and so the 3G personal privacy problems will persist into 4G.

In Chapter 7 a different way of handling identity presentation and entity authentication is presented. Here not only entity authentication is provided, but also credible subscriber privacy protection. Due to regulatory requirements the privacy protection must be revocable. For the privacy scheme to remain credible the revocation procedure should be such that neither the HE nor the SN can revoke the subscriber privacy alone.

In this chapter we look at various aspects of personal privacy for the mobile subscriber and investigate some of the proposed solutions.

5.2 Mobile Subscriber Privacy Aspects

5.2.1 The 3GPP Privacy Requirements

The 3GPP security architecture defines the following user identity confidentiality requirements [34, secion 5.1.1]:

- **User identity confidentiality:** the property that the permanent user identity (IMSI) of a user to whom a service is delivered cannot be eavesdropped on the radio access link.
- **User location confidentiality:** the property that the presence or the arrival of a user in a certain area cannot be determined by eavesdropping on the radio access link.
- **User untraceability:** the property that an intruder cannot deduce whether different services are delivered to the same user by eavesdropping on the radio access link.

The requirements are met, according to TS 33.102, as follows [34, section 5.1.1]:

> To achieve these objectives, the user is normally identified by a temporary identity by which he is known by the visited serving network. To avoid user traceability, which may lead to the compromise of user identity confidentiality, the user should not be identified for a long period by means of the same temporary identity. To achieve these security features, in addition it is required that any signalling or user data that might reveal the user's identity is ciphered on the radio access link.

> Clause 6.1 describes a mechanism that allows a user to be identified on the radio path by means of a temporary identity by which he is known in the visited serving network. This mechanism should normally be used to identify a user on the radio path in location update requests, service requests, detach requests, connection re-establishment requests, etc.

The 3GPP requirements are reasonable and necessary, but not comprehensive. The 3GPP security architecture does not capture the case were one requires identity/location privacy from cellular system entities. This is increasingly the case, since one cannot realistically expect the SN (there may be several hundred different SNs) to adhere to the HE/UE privacy policy. As discussed in Section 4.10.2 "The Enemy Within" one must also take into account that the SN and HE may try to willfully transgress on the subscriber rights. HE/SN tracking of subscriber position and movement, even if anonymized with respect to actual user identity, is at best unethical and at worst infringement of subscriber rights.[1] The *user untraceability* requirement is a complex one. It not only requires protection against tracking of identified subscribers, but it must also provide protection against cases where the adversary does not know the UE identity.

5.2.2 3GPP Subscriber Privacy Provision

The 3GPP privacy requirements, while not complete, are precise and well conceived. However, the 3GPP security architecture offers only a partial and somewhat inconsistent solution. The UMTS/LTE access signalling is schematically similar to the GSM access signalling and with a scheme that relies on symmetric crypto-primitives it is very hard to conceal the subscriber identity during the initial registration. One may use symmetric group keys to solve this problem, but there are serious problems with secure management of shared group keys.

The initial identity presentation must necessarily precede the identity verification (authentication). Thus, identity presentation takes place before the encryption keys have been agreed. Therefore, in the 3GPP scheme the permanent subscriber identity (IMSI) *must necessarily* be transferred in plaintext on the over-the-air interface. This is by and large unacceptable since it allows for subscriber location tracking. To mitigate the problem the Serving Network (SN) will normally issue a local temporary identity, TMSI (4 bytes wide), that is to be used for subsequent subscriber identification. The TMSI is assigned after the authentication and after the encryption (data confidentiality) has started. The subscriber and SN will use the TMSI, in plaintext, for system access and paging. This scheme will provide a certain level of subscriber anonymity since there is no apparent correlation between the ciphertext version of the TMSI (which is indirectly bounded to the IMSI) and the plaintext version of the TMSI. The would-be adversary is therefore unable to associate the IMSI and TMSI identifers.

However, the scheme is inadequate towards active attacks since the signalling procedures permits an SN network to simply request the IMSI identity. A false basestation would easily be able to collect identities. The identity privacy scheme also suffers from vulnerabilities that may allow passive attacks to succeed [39]. There is no formal requirement on exactly how the TMSI should be allocated. Some vendors have permitted the operators

---

[1] There are notable exceptions: Lawful Interception and Emergency Call functionality are the obvious ones. There is also the issue of location based services, which of course is acceptable provided explicit user consent has been obtained.

to add structure to the TMSI and it has been observed that some operators allocate the TMSI in a sequential fashion.[2] The fact that the allocation may be predictable can be used by an intruder to correlate the IMSI and the TMSI. For an active attack case it is also relatively easy to obtain the corresponding MSISDN number.

The topic of subscriber identity confidentiality was investigated during the UMTS design phase. The 3GPP SA3 work group had a work item called "Enhanced User Identity Confidentiality (EUIC)", which prior to the functional freeze of UMTS Release 99 was contained in TS 33.102 Annex B (Annex B is now *void*). The EUIC proposal used encrypted identities in the paging message (group key scheme) and it would require additional nodes. There would also be significant overhead in terms of processing and the setup delay would increase considerably (mostly due to the round-trips).[3] Retro-fitting functionality like subscriber privacy in a control plane model with mandatory UE identity presentation would never be easy, and there was considerable doubt that the proposed EUIC scheme would be effective. The EUIC proposal therefore never made it into the released security architecture.

5.2.3 Subscriber Privacy: Address vs. Identity

The subscriber privacy problem is rooted in the way identities and identity presentation are handled in the cellular systems. To improve on the subscriber privacy one must closely examine the whole access setup, which would include investigating the identity presentation, registration/location handling (local and global) and the subscriber information transfer (subscriber service context data) and finally the security setup (the AKA protocol).

The subscriber privacy problem is directly related to the fact that the IMSI is not only a subscriber identity, but also an address to the HLR/AuC at the HE. An identity is an abstract concept while an address is a concrete way of locating a resource. In cellular systems the IMSI must serve both purposes. During initial registration the SN needs to have an address to the HE. The only way to do this in the cellular system architecture is to use the IMSI to derive

---

[2] One scheme involved subdividing the TMSI state space such one octet was allocated to identify the VLR/MSC. The remaining 3 octets were sequentially allocated (out of a pool of available TMSI identities) by the VLR/MSC.

[3] The proposal, contained in temporary document S3-99459_CR33102-22r1.rtf, is still available at the 3GPP web site (http://www.3gpp.org/ftp/tsg_sa/WG3_Security/TSGS3_08/Docs/S3-99459_CR33102-22r1.rtf).

an address to the HE[4]. The SN *must* therefore know the IMSI. A separation of identification and address resolution should be sough for a future system such that the subscriber identity can be kept private if necessary.

5.2.4 Practical Limits to Identity Privacy

In the PE3WAKA protocols[5] one needs to protect the UE identity from eavesdropping by external parties. One also wants to protect the privacy of the UE with respect to the SN and HE. As will be discussed in the subsequent sections (Section 5.4.1), it is possible to prevent the SN from knowing the long-term identity of the UE and it is possible to prevent the HE from knowing the UE location.

In [130] Krawczyk discusses the so-called SIGMA class of protocols (see also Section 5.5.6 for a brief discussion). The SIGMA protocols are versatile protocols and they are capable of providing identity privacy. As is noted in [130] there is a limit to the extent of identity privacy one can achieve. Specifically, it is noted that one cannot provide identity privacy to protect both peers from active attacks:

> As it turns out the requirement to support identity protection adds new subtleties to the design of KE protocols; these subtleties arise from the conflicting nature of identity protection and authentication. In particular, it is not possible to design a protocol that will protect both peer identities from active attacks. This is easy to see by noting that the first peer to authenticate itself (i.e. to prove its identity to the other party) must disclose its identity to the other party before it can verify the identity of the latter. Therefore the identity of the first-authenticating peer cannot be protected against an active attacker. In other words, KE protocols may protect both identities from passive attacks and may, at best, protect the identity of one of the peers from disclosure against an active attacker. (From section 2.2 in [130])

With respect to the PE3WAKA protocols it is noted that there is no requirement to confidentiality protect the identities of the SN and the HE. There is therefore no conflict between the observations in [130] and the requirements for the PE3WAKA protocols.

---

[4] The derived address is an ITU-T E.214 [129] Mobile Global Title (MGT) type of address.

[5] The PE3WAKA (Privacy Enhanced 3-Way Authentication and Key Agreement) acronym is used for the protocol family presented in Chapter 7.

Table 5.1 Knowledge of UE privacy sensitive data by domain.

Privacy Dimension	HE knowledge	SN knowledge	UE knowledge
UE Location	*No*	Yes	Yes
UE Identity	Yes	*No*	Yes

5.3 Regulatory Privacy Limitations

As was presented in Section 4.5.4 public cellular systems are subject to regulatory requirements. These include emergency call requirements and lawful interception requirements. Recently, data retention requirements have also emerged. The regulatory requirements mean that the subscriber privacy must be revocable. Ideally, the revocation should technically be a non-trivial procedure such that no single entity can revoke the subscriber privacy. That is, given that the appropriate regulatory data access procedure is followed, it would still be preferable that the revocation scheme requires both the HE and SN to cooperate for the subscriber data to be revealed.

5.4 Cellular Privacy Principles

5.4.1 Domain Separation

The HE assigns the long-term UE identity ($UEID$). Thus, UE identity privacy from the HE is not possible. Conversely, the SN will necessarily be able to know the approximate UE position.[6] Thus, UE location privacy from the SN is not possible. See also the discussion in Section 4.6.

Table 5.1 gives an overview over *necessarily possible* knowledge of privacy sensitive subscriber identity and location information. By *necessarily possible* knowledge one means that the principal may obtain the information, and that the other principals cannot deny the principal from obtaining this knowledge. However, Table 5.1 also indicates that one may provide *UE Identity Privacy* from SN monitoring and *UE Location Privacy* from HE monitoring. To attain subscriber privacy one must ensure that the control plane signalling does not unnecessarily distribute subscriber privacy sensitive information. The identity management scheme and PE3WAKA protocol must therefore guarantee that the HE cannot learn the UE location and that the SN will not learn the permanent UE identity.

The subscriber may choose to divulge the UE position to the HE or other parties to obtain location based services. This is an application layer decision

---

[6] Strictly speaking, it will be the MT position that the SN will know.

and it is a different matter from leakage of privacy sensitive data at the lower layers. The idea of knowledge separation based on domain is not new. Similar requirements and arguments are also found in [15].

5.4.2 Identity Management and Subscriber Privacy

The Subscriber Privacy Problem

The subscriber privacy problem can be formulated as follows:

- **UE Location**
  The UE can allow the SN to know its position, but the HE should not normally be able to learn the UE position. Entities that observe the A-interface may be able to establish the UE location, but these entities should not be able learn which UE they observe.
- **UE Identity**
  The UE can allow the HE to know its permanent identity, but the SN should not normally be able to learn the permanent UE identity. Entities that observe the A-interface should not be able learn the UE identity.
- **Traceability**
  Neither the SN nor the HE should normally be able to track the UE over longer periods of time. No other entity should be able to track the UE over longer periods of time. The tracking requirements also apply to tracking of anonymous UEs.

The regulatory requirements, as outlined in Section 4.5.4, still applies.

Outline of a Solution

To solve the problem with subscriber identity and location privacy a random reference identity scheme is proposed. In Section 4.8.4 it was shown that the setup performance would be improved by combining and integrating the *identity presentation*, the *authentication and key agreement* and the *registration* procedures. For this to work it is necessary to let the UD/UE be the initiator of the combined procedure, since it is the UD that is responsible for detecting AN/SN network changes, like new area codes, etc.

To provide subscriber privacy it is essential that the UD/UE is not forced to provide identity information in plaintext over the A-interface. It is also essential that the UE does provide its identity to the HE. The SN does not need to know the permanent UE identity, but it is entitled to assurance from the HE that it will accept liability for the charges incurred by the UE. To this end it is necessary that the UE, SN and HE have a common reference to the UE. This common reference should have limited temporal validity, or else it would simply become a permanent alias identity for UE. One way of containing the UE reference identity is to bind it to the roaming context (similar to the $MTC_{(HE,SN,UE)}$ security context).

The Random Reference Identity Scheme

The following is an outline of a random reference identity scheme. The random reference identity, which shall be denoted simply as the *context identity* (CID), is a temporary subscriber (UE) identity and acts as a common reference for the roaming context. The roaming context is equivalent to the shared *medium-term context* ($MTC_{(HE,SN,UE)}$) which was defined in Section 4.7.4.

Location privacy from SN monitoring hinges on the SN not knowing the permanent UE identity. One must therefore conceal the permanent UE identity ($UEID$) from the SN. To do so has some ramifications: (a) the UE identity cannot be used by the SN as an address to the HE and (b) the SN will need some type of temporary identify to keep as a reference to the roaming context. The latter problem is solved by letting the context reference identity (CID) be unique (in the context) and recognized by all three principals.

The registration procedure is initiated by the UE. During the initial message ($M1$) from the UE to the SN enough information must be given for the SN to be able to contact the HE. That is, the message must contain the HE identity ($HEID$) or an HE address. The $M1$ message must also contain the $UEID$ and it must be confidentiality protected such that only the HE can decipher it. The only viable solution here is to use an asymmetric encryption method. The UE must therefore encrypt its identity (the $UEID$) with a public key belonging to the HE. Since the HE may have multiple public-key key pairs the $M1$ message may also need to include a reference to the key pair (or the digital certificate).

When the UE decides to execute the *registration* procedure it will choose the CID identity. The CID must be constructed such that there is no apparent correlation between the permanent identity $UEID$ and the CID identity. It is therefore proposed to let the UE generate the CID value by means of a pseudo-random function. The CID will be the common reference to the $MTC_{(HE,SN,UE)}$ security context. The CID must be included in message $M1$. The CID should preferably be a personal privacy secret only known to the UE, SN and HE, but this requirement may be relaxed under some circumstances.

Given a regime with randomly generated identities there will be a certain risk of identity collision. The collision events will be relative to the SN.[7] Collision events are undesirable, and the SN must reject access attempts with a duplicate CID. To reduce the probability of a CID collision one must ensure that the CID information element has a sufficient state space. If one assumes that there is no bias to the CID choices one may use the approximation $p = k/m$, where p is collision probability, k the maximum number of users within the SN server area and m is the state space of the CID. To be on the safe side let us assume that an SN domain will have at most ten billion simultaneous users ($k = 10^{10}$). If one very conservatively requires collision events to occur for at most every 100 million AKA occurrence ($p = 1/10^8$) then the CID must have a state space of minimum $m = 10^{18}$.

These requirements indicates that the CID information element needs to be approximately 60-bit wide ($2^{60} > 10^{18}$). This is clearly not a problem. It is therefore concluded that it is viable for the UE to generate the CID by a suitable cryptographic pseudo-random function (prf).

The provision of a pseudo-random identity does solve our problem with identity privacy for the PE3WAKA protocols, and the use of asymmetric encryption makes sure that the attained identity privacy is effective against both external adversaries and the SN. But the scheme is insufficient with respect to subscriber tracking protection. It is therefore proposed to introduce yet another anonymous UE identity. The identity is a session identity (SID), and this identity will only be used locally between the UE and the SN. The SID will be associated with the $STC_{(SN,UE)}$ context, and as such it will be used in plaintext by the UE and SN over shared channels during paging and UE access requests. The UE and the SN will maintain an association between the CID and the SID for the duration of the $STC_{(UE,SN)}$ context. The HE need not know the SID. The SID may be generated by the UE, but to ensure uniqueness it is better if the SN allocates the SID. This also provides a measure of balance with respect to identity assignment. This will allow improved initiator-responder properties for the derivation of key material, which should include the context/session identities as input for proper context binding.

The above sketched use of SID does not fully prevent subscriber tracking since the SID may be used multiple times during the (UE,SN) session. However, one may consider the privacy goal to be satisfied provided that

---

[7] When generating the CID the UE need only guarantee uniqueness with respect to the $(CID, UEID, SN)$ tuple. This will ensure that the UE and HE will not experience identity collisions. For the SN, which will not know the $UEID$, the CID must be unique across all visiting UEs.

the (UE,SN) sessions are sufficiently short-lived. While one may limit the duration of a lower layer session one cannot reasonably require the higher-layer user sessions to be short-lived. A user session may therefore outlive the $STC_{(SN,UE))}$ context. For a packet-switched architecture this need not be problematic, but it is observed that the SN will be able to track high-layer sessions spanning several $STC_{(SN,UE)}$ context. The SN tracking capabilities would then at least span the lifetime of the $MTC_{(HE,SN,UE)}$ context, and due to IP header information leakage, etc., the SN may also be able to track the UE for multiple-$MTC_{(HE,SN,UE)}$ contexts. The SN may therefore be able to correlate a sequence of CID identities. Tracking by the SN, which after all is a semi-trusted entity, is less of a worry than if external entities are able to track the UE.

It is noted that from an information theoretic perspective one must prove the absence of any kind of correlation between each access (including usage patterns, etc.) to prevent tracking. This would include means to de-correlate physical radio related information.

The above outline and the justification for tracking protection with random session identities is made with the assumption that the system uses common channels for paging requests and access requests. The assumption holds true for the current cellular systems, but it may not necessarily be true for a future mobile system. In fact, it is easy to envision systems where paging and access requests take place on subscriber specific control channels. For ultra-wideband systems with OFDM/OFDMA scheduling one may for instance permanently allocate a subscriber specific (narrow-band) control channel as part of an integrated AKA protocol. Then there will be no need for session identities since the session oriented control signalling will be conducted on private secured channels. Instead, the problem will be to de-correlate the channel association since it will now be possible to track a subscriber by tracking the usage of specific and identifiable control channels. This problem may be more difficult to solve, but it may be possible to solve it using "pseudo-random channel hopping" or other similar methods.

The "random reference identity scheme" outlined here forms the basis for subscriber identity privacy in the PE3WAKA protocols. The PE3WAKA protocol family is further described in Section 7.4.

5.5 Other Privacy Technologies

5.5.1 Overview over Related Research and Alternative Solutions

There have been many proposals on how to provide and handle subscriber privacy for mobile systems. The discussions in some of these papers are interesting [11–14], but the solutions are not particularly well fitted to meet the needs of future mobile systems.

An alternative scheme using group pseudonyms is investigated in [131]. This scheme is GSM specific and not particularly relevant for a future cellular system. In [11] the authors investigate the use of MIXes. However, MIXes typically have non-deterministic delays and one would need a set of MIXes to achieve sufficient privacy.

In the papers [12, 15] the subscriber privacy requirements is captured quite well. However, the solutions presented do not offer sufficient home control. In fact, it was a design goal to avoid the global round-trip back to the HE. While the HE does not necessarily need to be online for all procedures, there is nevertheless a strong requirement for HE Home Control and in practice this requires the control protocol to be an online protocol. The security context hierarchy, as outlined in Section 4.7.4, is very useful in that it can provide a flexible scheme that will permit improved home control and subscriber privacy.

The PE3WAKA scheme is also very different from [12] when it comes to identity management. Another paper that deals with user privacy and wireless authentication is [13]. Again, we find that many of the requirements are similar to our requirements, but proposed solutions are quite different from the PE3WAKA approach. A direct comparison between the PE3WAKA protocols and the suggested protocols in [13] would be unfair since they deliberately considered low-cost solutions while we aim at a complete solution with a redesigned identity scheme. Finally, many of the requirements we have arrived at are also found in [14]. The solution suggested in [14] does not take into account the possibility of integrating security setup with the mobility management procedures, and it has a very different approach to identity management.

The SIGMA class of protocols [130] optionally provides identity privacy for the initiating principal. The SIGMA class of protocols is therefore of interest, and it is instructive to see how the SIGMA protocols combine asymmetric and symmetric methods to achieve their goals. However, the SIGMA protocols are two-party protocols and focus on key exchange rather than on entity authentication. The SIGMA protocols, while they may serve as a source of inspiration, do not quite fit the PE3WAKA environment.

5.5.2 Secure Multi-Party Computation

Is it possible to simultaneously provide enhanced home control and a credible subscriber privacy scheme? That is, is it possible to let the HE verify that the UE is inside/outside some geographical region and still avoid disclosing the UE position to the HE? The requirements may seem irreconcilable, but it is possible to achieve this using Secure Multi-part Computation (SMC) methods.

The Point Inclusion Problem

The basis for solving the problem with enhanced home control and subscriber location privacy is the Secure Multi-Party (SMC) problem known as the *point inclusion problem* [132, 133]. It can be defined as follows: Alice has a point z and Bob has a polygon P. The problem is to determine whether z is located within the polygon P such that Alice does not have to disclose any information about the point z to Bob or that Bob does not have to disclose information on the polygon P to Alice. The only fact to be revealed through the procedure is the answer to the problem. The problem is a special case of a general secure multi-party computation problem and it is known that it can be solved by using a circuit evaluation protocol [134]. However these solutions are impractical because of high communication complexity. Therefore special solutions of the problem have been proposed in the literature [132, 133, 135]. They are more efficient, but are nevertheless impractical for use in a cellular system.

Suitability of SMC in a Cellular Context

An optimized 2-party point inclusion protocol was therefore devised. This protocol, known as the *Secure Two-Party Location Inclusion Protocol* (S2PLIP), is described in [6, 8]. However, the S2PLIP is still a complex protocol and it must be integrated with the PE3WAKA protocol. The overall computational and communication complexity may be too high for S2PLIP to be a practical proposal. As was demonstrated in [8] the S2PLIP can be integrated with a PE3WAKA protocol such that the round-trip number does not increase from the basic PE3WAKA protocol, but the computational and communication overhead is nevertheless problematic. One may reduce the overhead by having simple polygons, but the S2PLIP protocol will still be a demanding protocol to run during context establishment. Having evaluated the S2PLIP protocol, it was decided that this type of privacy-preserving spatial verification is best done subsequent to AKA protocol execution. The S2PLIP protocol has therefore not been integrated with the AKA protocols described in Section 7.4.

5.5.3 Pseudonyms

Pseudonym and Symmetric Group Keys

The 2G/3G scheme of using a temporary identity (TMSI) in place of the permanent identity (IMSI) is of course a pseudonym scheme in which the TMSI is the pseudonym. However, as has been discussed, the TMSI scheme is not able to provide credible privacy.

The 3GPP *Enhanced User Identity Confidentiality* proposal was based on a group key scheme (see Section 5.2.2). The proposal in effect transformed, by an encryption transform, the long-term identity into a short-lived pseudonym identity. A basic problem with the EUIC scheme was the insecurity associated with shared symmetric keys and the fact that the 3GPP access signalling scheme could not easily accommodate encrypted identities. That is, while the EUIC scheme may have been functionally feasible it required additional round-trips[8] and it required new nodes and new functionality in the networks. The processing requirements were in fact modest, but there was very little willingness in 3GPP to accommodate schemes that required changes to the signalling flow and message expansion (which in turn may have required message segmentation) since such a redesign would likely have delayed the overall system design.

Other group pseudonym schemes have also been proposed. In [131] the authors propose several schemes that use pseudo-random number generators, hash functions and public-key signatures. The proposals were directly targeted at the GSM system access signalling (and would by-and-large be compatible with UMTS access signalling). The proposals in [131] were however *much* more complicated that the EUIC scheme and would probably be unsuitable for access signalling.

---

[8] There were various suggestions to mitigate this, but none were seen to solve the issues completely.

5.5.4 MIXes

MIX networks

The MIX idea was first proposed by Chaum in [10]. The MIX network anonymizes addresses and provides protection against traffic flow analysis. The MIXing involves message padding, message and address encryption and it can include injecting random delays in the transfers. As such the MIX networks are therefore unsuitable for real-time processing and furthermore the MIX networks assume that the all MIX network nodes can be trusted to carry out the MIXing operations. Thus, the users of the MIX network must trust it to operate as promised with respect to the actual obtained privacy.

MIXes in Mobile Networks

MIX networks, in the general case, are not suitable for real-time communication, but one may tailor the MIX networks to the task at hand. For non-time critical asynchronous communication the non-deterministic delays induced by the MIX network may not matter, but the control plane signalling is almost always time critical. For instance, almost all of the mobility management handling is trigged by physical events and the required response time is measured in milliseconds (or less).

In [11] the authors propose to introduce as set of dedicated MIXes. The suggestions in the paper are directly aimed at a GSM target architecture and it was proposed to let the BTS be the MIX termination point. The reason for this is that the MIXes expand the data and data expansion on the air-interface could not be tolerated in GSM. More modern mobile systems have improved over-the-air capacity, but the over-the-air channel is nevertheless a capacity bottleneck and the advice to avoid message expansion on the A-interface is sound. The same paper [11] also advocates placing the MIXes close to the HE/SN network nodes to avoid additional propagation delays. This is good advice since the signalling procedures (mobility management) are very delay sensitive. In [136] it is noted that timed MIXes are vulnerable to attacks and the authors propose to use a cascade of MIXes with batch/synchronization scheme. The proposal in [136] is targeted for WLAN networks. The authors seem to be aware that the MIX network configuration (called the defense protocol WLP^2S in the paper) is not suitable for real-time processing.

The problem with the MIX solution as outline in [11, 136] is that the MIXes, whatever the location, will induce additional delays and management overhead. It is also difficult to see why the HE and SN would want to deploy and operate a dedicated cellular system MIX network. Even worse, as has already been noted, the subscriber privacy in a public cellular system is inevitably subject to public regulatory control and the privacy must be revocable. This is not to say that a MIX network approach will not work as intended, but it does not seem like an optimal solution. It is therefore concluded that the MIX network approach is not well suited to provide subscriber privacy for mobile system access signalling cases.

5.5.5 Privacy Trusted Third Party

Technically the use of a trusted third party as subscriber privacy proxy is clearly feasible. From a performance point of view the inclusion of a privacy-TTP between the HE and SN networks is only a minor inconvenience, and the performance penalty need not be excessive. As for the MIX network case, the solutions *may* be vulnerable to traffic flow analysis for the real-time case, but this may be tolerable.

From a trust perspective the HE, SN and UE must obviously all trust the privacy-TTP. From an operational point of view it is difficult to see the incentives to operate a high-availability real-time privacy-TTP. The TTP could be operated by a regulatory actor. However, despite the fact that these actors *may* obtain legal access to the data, it seems unwise to permit a regulatory actor to obtain the privacy sensitive information by default.

5.5.6 The SIGMA Approach

The SIGMA class of protocols [130] provides identity privacy for the initiating principal. The SIGMA class of protocols is therefore of interest, and it is instructive to see how the SIGMA protocols combine asymmetric and symmetric methods to achieve its goals. Figure 5.1 illustrates the basic SIGMA scheme with identity protection.

The SIGMA protocols are two-party protocols and the primary focus is on key exchange rather than on entity authentication. The key exchange (or rather key agreement) is through a DH exchange. It is noted that the Internet Key Exchange (IKE) protocols (version 1 [137] and version 2 [138]) belong to the SIGMA class of protocols. The SIGMA protocols, while the convoluted design does serve as a source of inspiration, is inadequate for the 3-way cellular access security setting.

The keys K_e, K_m and the session keyset K_s (not shown) are all derived from the same DH secret. The keys must be computationally independent for

(M1) A → B: g^x
(M2) B → A: $g^y, \{B, SIG_B(g^x, g^y), MAC_{K_m}(B)\}_{K_e}$
(M3) A → B: $\{A, SIG_A(g^y, g^x), MAC_{K_m}(A)\}_{K_e}$

The Identifiers:

A, B:	The principal entities
g^x, g^y:	The DH exponentials from A and B respectively
K_e:	Symmetric cipher key for confidentiality (derived from DH exchange)
K_m:	Symmetric cipher key for message integrity and origin authentication (derived from DH exchange)

Figure 5.1 Simplified SIGMA protocol with identity protection.

the scheme to be secure.[9] The basis for the public-key signatures (SIG_B and SIG_A) must be agreed before executing the protocol.

5.6 Discussion and Summary

5.6.1 Discussion

A crucial issue with the subscriber privacy solutions is whether they are both credible and practical. The proposed scheme with domain separation and use of random (anonymous) identities has several strengths and does appear to be a very good solution.

The requirement that the subscriber privacy must be revocable given the proper legal authorization is troublesome. However, the revocation requirement must be considered an absolute requirement and it serves no purpose to debate the requirement as such. Revocable personal privacy is therefore a necessary feature. And in this context, the proposed scheme has the redeeming characteristic that no single entity can execute the revocation. This is a safeguard against abuse of the privacy revocation feature.[10]

The proposed random identity scheme is based on a set of premises. The validity of the premises is crucial here, and they must therefore be closely examined. A central premise for subscriber privacy is that of domain separation, which ensures that each party (SN and HE) only learns what it necessarily must know. The issue rests with the assumption that the SN and HE domains

---

[9] That is, knowledge of a key X1 does not aid in finding another key X2.

[10] The safeguard is arguably weak, but when seen as measure for defense in depth it still has value.

are truly independent in terms of management control. The fact is that this assumption is not necessarily true.

As is noted that operators of a SN/AN network will in likelihood also have HE functionality. The SN/AN is then referred to as the Home Network (HN). In terms of management control, the operator of a HN may obviously violate the domain separation assumption. Thus, the HN operator can correlate the subscriber information and effectively revoke the subscriber privacy. The HN is the primary subscription area and the subscriber is expected to spend most of his/her time within this area. Does this mean that the domain separation scheme is based on a invalid premise and that it cannot in reality provide subscriber privacy? In theory, this is a devastating blow to the domain separation scheme since one apparently cannot be sure that the domains are separated. In practice, the question is more complex. The fact that something can be done does not mean that it has to be done. To believe that the HN will not correlate the subscriber identity/location information does require a higher level of UE trust in the HN. This may not always be warranted. One way of fixing this problem is to have regulatory requirements to enforce domain separation. That is, regulatory enforcement to separate the domains in terms of exchanged information. This is not only possible, but to some extent this type of regulatory regime is already in operation in many countries in the 2G/3G networks. The reason has nothing to do with subscriber privacy *per se*, but is a necessity in order to provide virtual HEs (vHE) fair access to SN/AN networks.

One must also note that the fact that the HE and SN domains are under control of distinct and separate administrative entities does not in itself preclude them from exchanging the privacy sensitive information. A question that must be asked then is if there are alternative subscriber privacy methods that avoid the problems with domain separation. That is, can one avoid the whole issue of domain separation in a public cellular system and still have credible subscriber privacy? The short answer is *no*. The reason is simple. Most regulatory authorities no longer permit anonymous subscriptions. That is, the HE operator is required to know the identity of the legal entity that subscribes to the service (and thereby creating the link between the UE identity and the legal entity). Then, as outlined in Section 5.4.1 and Table 5.1, one ends up with the case that the SN must necessarily know the approximate MT location and the HE must know the UE identity and the identity of the associated legal entity.

One may argue that intermediate privacy-TTP nodes and/or MIX networks may solve the issue, but as was argued in section 5.5 these solutions are generally impractical in the cellular access context. The MIX and Privacy-TTP solutions themselves also introduce trust issues, and it is not at all obvious that the UE, SN and HE would want to trust a MIX network or a Privacy-TTP.

5.7 Summary

In this chapter subscriber privacy, with focus on *identity privacy* and *location privacy*, is investigated. The UMTS subscriber privacy requirements were examined and the requirements were found to be reasonable but incomplete. The UMTS privacy scheme, which was inherited from the GSM system, does not however do a very good job at meeting the requirements. In order to attain improved subscriber privacy a new identity management model is proposed. The model is based on the security context hierarchy found in the previous chapter, and involves changing the initiator-responder model to better match the triggering events for system access. The identity management involves three layers; one layer for the permanent subscriber identity, one layer for a medium-term reference identity and one layer for a session identity. The "random reference identity scheme" (Section 5.4.2) forms the basis for the AKA protocols devised in Section 7.4.

The proposed scheme rests on the assumption that the HE and SN domains are separated with respect to subscriber privacy information. Alternative schemes were investigated, but were found not to be suitable for the cellular environment.

6

Principles for Cellular Access Security

> The promise given was a necessity of the past:
> the word broken is a necessity of the present.
>
> *– Niccolo Machiavelli (1469–1527)*

6.1 Introduction

This chapter is intended to sum up the requirements and principles derived in Chapters 4 and 5. The approach is generic and no specific cellular system architecture is presented. A system model was presented in Chapter 4, but the given model is still quite generic. So instead of presenting a set of concrete design requirements for the access security architecture the requirements have been formulated as a set of design principles.

6.2 Generic Security Protocol Design Principles

6.2.1 The Guidelines Approach to Security Protocol Design

How should one design a security architecture for a mobile system? There is no definitive answer, but we know from experience how *not* to design security protocols and security architectures. In their 1994 DEC System Research Center report "Prudent Engineering Practice for Cryptographic Protocols" [139] Abadi and Needham defined a set of engineering principles for designing security protocols. The advice and guidelines in [139] are largely derived from actual errors and faults found in proposed security protocols.

The requirements for the 3G/UMTS security architecture also provides us with design input [38]. With the principles and 3G/UMTS requirements in mind, we define additional requirements and guidelines for the design of future mobile security protocols. The principles provided by Abadi and Needham is not the only set of guidelines proposed. In fact, there are quite a few papers that recommend the use of design principles and design guidelines. A follow-up paper to the Abadi–Needham paper is presented in [140]. This paper provides robustness principles for public-key systems and extends [139]. In "Systematic Design of a Family of Attack-Resistant Authentication Protocols" [141] the authors provide advice on how to construct security protocols.

6.2.2 The Prudent Engineering Principles

The Abadi–Needham prudent engineering principles for security protocols are summarized below. A full account of the principles is in DEC SRC report no. 125 [139]. The principles represent sound engineering practices and one should have compelling arguments if one decides to violate a principle. There is an additional unacknowledged principle. This principle states that one cannot trust unprotected information. This principle, which should be self-evident, is expressed as principle 0 and its origin can be traced to the influential DEC SRC report no. 39 "A Logic of Authentication" (the BAN logic report) [66]. Given that one cannot trust unprotected information it makes sense to propose to protect all information that needs to be protected. This is captured in the second (corollary) part of principle 0.

The Prudent Engineering (PE) principles are seen as a useful starting point and they are accepted as a base. However, the PE principles are not domain specific and they do not account to environmental considerations. Thus, the PE principles must be augmented to be truly useful as a design method tool for the PE3WAKA protocol family. The PE principles will be referred to as **PE.X**, where X is the principle number. For example **PE.10** refers to the *Trust relationships should be explicit ...* principle.

The Abadi–Needham Prudent Engineering principles from [139] (adapted):

0. Unprotected information cannot be trusted and cannot be used to develop security beliefs (the term "belief" has a special meaning in BAN logic).

   Corollary:
   All information that is to be trusted must be protected. Secret information must be confidentiality protected and all trusted information must be integrity protected.

1. Every message should say what it means. The interpretation of the message should depend only on its contents.
2. The conditions for a message to be acted upon should be clearly and unambiguously defined. Someone reviewing the design should immediately see if the conditions are acceptable or not.
3. If the identity of the principal is essential to the meaning of the message, then the principal name should be explicitly given in the message.
4. Be clear about why encryption/cryptographic methods are being used. (That is, which security service is it that they provide.)
5. When a principal signs information elements that is already encrypted, it should not be inferred that the principal knows the encrypted contents.
6. Be clear about why nonces are being used
7. If a predictable quantity (such as a counter) is used to guarantee newness, then it must be protected to prevent an intruder from simulating the sequence.
8. If timestamps in absolute time are used as nonces, then the difference between local clocks must be taken into account (both to limit the "replay-window" and to prevent premature expiry). The integrity of the system clocks must be ensured and synchronization between the clocks must be protected.
9. It should be possible to deduce the protocol, which run of the protocol and the message number within the sequence by inspecting a message.
   Extended interpretation: The protocol state machine must be fully defined. The messages must be completely specified and the encoding of all information elements must fully defined.
10. Trust relationships should be explicit. The reason for the relationships should be clear.

6.2.3 High-Level Practical Guidelines

To derive guidelines for security protocols is rarely an end in itself. Still, when designing authentication and key distribution/agreement protocols one should have defined high-level goals for the new protocols. These goals should be explicit and should provide clear and verifiable advice on how to construct the protocol.

The IPsec protocol suite is now in its third version and during the design and development of the third version there was a heated debate on how to proceed with the design of key distribution in IPsec. In particular, one fraction within the IPsec working group favored a simple and easy to understand approach. They named their proposal protocol Just Fast Keying (JFK), and it was aimed at providing fast and reliable key distribution for the IPsec protocol suite. The JFK protocol was not to provide complete functionality, but was rather designed to solve the main key distribution problem for most of the users. This would allow the JFK protocol to be a relatively simple protocol and JFK could therefore be a fast, efficient and reliable protocol. Users that would require extended services would have to resort to IKE version 1 (RFC 2409, etc.) or other protocols to meet their requirements. The design goals of JFK is stated in [142, section 1.1]. In particular, the design goal for the JFK protocol was to provide (quote from [142]):

- *Security:* No one other than the participants may have access to the generated key.
- *PFS:* It must approach Perfect Forward Secrecy.
- *Privacy:* It must preserve the privacy of the initiator and/or responder, insofar as possible.
- *Memory-DoS:* It must resist memory exhaustion attacks.
- *Computation-DoS:* It must resist CPU exhaustion attacks on the responder.
- *Efficiency:* It must be efficient with respect to computation, bandwidth, and number of rounds.
- *Non-Negotiated:* It must avoid complex negotiations over capabilities.
- *Simplicity:* The resulting protocol must be as simple as possible, within the constraints of the requirements.

The JFK design goals have not been elevated to design principles or included in the guidelines, but they are nevertheless worth noting as the goals are sensible and well conceived. As a matter of fact, several of the JFK goals are reflected in the design principles for the cellular access security guidelines.

6.3 Cellular Security and Subscriber Privacy Principles

6.3.1 Cellular Security Requirements and Guidelines

The Abadi–Needham principles are not sufficient for the design of the PE3WAKA protocol family and so additional requirements and guidelines for the cellular system environment are added. The fact that the protocol environment plays an important role is not always acknowledged in the literature, but in the paper "Environmental Requirements for Authentication Protocols" [143] the topic is thoroughly investigated. It is contended that security is a system property, and that security must be designed in a system context.

Thus, one *must* address aspects like the system service model, the system control model, acceptable subscriber perceived performance etc. Performance is a critical success factor and one really cannot afford to let the security protocols significantly degrade system performance. Performance has many aspects. For the PE3WAKA protocols one must be particularly concerned with the temporal performance, but one still cannot ignore the computational and communications performance of the protocols. The guidelines/principles presented here are based on the analysis found in Chapters 4 and 5.

6.3.2 The Cellular Access Security and Privacy Guidelines

The following high-level requirements for an access security architecture have been identified. The requirements may not always apply, and the designer should judge their applicability and be selective when necessary. Adherence to these principles and guidelines is no substitute for formal verification. The guidelines will be referred to as **G.X**, where X is the principle number. For example **G.18** refers to the *Subscriber Privacy Override* principle.

High-Level Access Security Design Requirements and Guidelines

1. *Home Control*
   The HE is entitled to have a degree of enforceable Home Control. Arguments for home control is found in Sections 4.6 and 4.7.1.
2. *Online Security Context Establishment*
   The security context must be established online in order to guarantee freshness, and thereby providing real-time home control for the mobility management registration event. Arguments for online security context can be found in Sections 4.6 and 4.7.2.
3. *Active 3-Way Security Context Establishment*
   All principal parties must be *active* participants in the establishment of the security context; all principals should have a measure of influence on the established security context. Arguments supporting an active 3-way security context can be found in Sections 4.6, 4.7.2 and 4.7.3.
4. *3-Way Registration Control*
   In a cellular system the UE *registration* with an SN is a very important event. To ensure correct registration it is necessary to require active participation from both the UE and SN during the registration event. The HE must invariably participate in the registration due to the cellular mobility model, and so the registration procedure must therefore be executed such

that it takes a 3-way authenticated consent (HE,SN,UE) to complete the registration. This principle is implied from principles **G.1**, **G.2** and **G.3**. It is supported by the arguments for the above mentioned principles.

5. *Security Context Hierarchy*
To provide necessary performance and flexibility, and to provide integration with the mobile system topology, the security architecture must provide a security context hierarchy. The security context hierarchy must reflect the trust assumptions. Arguments supporting a security context hierarchy can be found throughout Section 4.7.

6. *Security Context Exposure Control*
The mobile security context must have sufficient exposure control. The control includes expiry conditions for usage, time and area. Arguments supporting a security context exposure control can be found in Section 4.7.4.

7. *Security Context Separation*
Short-term security contexts should never be shared over different access networks. Medium-term security contexts *may* be shared over different access networks, but should probably not be shared over different access types. Arguments supporting security context separation can be found in Section 4.7.4.

8. *Spatio-Temporal Security Context Binding*
The spatial and temporal control dimensions should be bounded to the security context explicitly. Usage based expiry will additionally be needed for session key expiry. Arguments supporting a security context binding can be found in Section 4.7.4.

9. *Security Algorithm Separation*
Key material in a (short-term) security context should never be used for more than one algorithm. This even extends to use of different modes-of-operation for the same algorithm. Arguments supporting security algorithm separation and proper key binding can be found in Section 4.7.4.

10. *Key Sets and Key Strength*
Both confidentiality and integrity protection is needed at the link layer. The protection is assumed to be provided by symmetric-key algorithms. It is assumed that the services are realized through independent cryptographic algorithms and then two independent 128-bit keys are needed. If a combined method is used then one 128-bit key may suffice. It *may* be best to separate the *control plane* and *user plane* data into separate security contexts. This is particularly true when the security for the *con-*

trol plane and *user plane* data terminates at different nodes. Arguments supporting context separation can be found in Section 4.9.2. The key sets may need to be directional. In particular, this may be the case for asymmetric user plane traffic provisioning.

Generally, the cryptographic strength should be of "128-bit quality".

The exception to the rule is for the integrity function in the cases where *collision resistance* is not an essential property. Arguments supporting "128-bit security" can be found in Section 4.9.

11. *SN Security Termination*

    Access security should terminate in the Access Network. The 3-way AKA protocol should terminate in a SN node in the core network. It is noted that *control plane* data and *user plane* data *may* be terminated at separate nodes.

12. *Identity Management, Registration and Security Setup*

    The procedures for Identity Management, Registration and Security Setup are logically connected and should be designed in conjunction. Arguments supporting a combined context establishment procedure can be found in Section 4.8.4. This principle is supported by principle **G.4**.

13. *Channel Allocation and Session Key Agreement*

    The procedures for Channel Allocation and Session Key Agreement may be executed in conjunction and should be designed to allow this.[1] Arguments supporting a combined session context establishment procedure can be found in Section 4.8.4.

14. *Subscriber Privacy – Domain Separation*

    Domain separation is the key to achieving subscriber privacy. The SN/AN and HE must therefore be administered and managed separately. Arguments supporting domain separation are found in Section 5.4.1.

15. *Subscriber Privacy – Identity Privacy*

    Only the HE and the UE should ever be allowed to learn the permanent UE identity. This is captured in Section 5.4.

16. *Subscriber Privacy – Location Privacy*

    Unless explicitly required; the HE should never be allowed to learn the UE location. This is captured in Section 5.4.

17. *Subscriber Privacy – Untraceability*

    No entity should be allowed to monitor UE movement over time. This is captured in Section 5.4.

18. *Subscriber Privacy Override*

---

[1] This will be for establishing the short-term session-oriented $STC_{(SN,UE)}$ context.

For better or worse, regulatory requirements may dictate that the UE position and identity be revealed. Given HE and SN cooperation it must therefore be possible to extract UE identity and position. This is discussed in Section 4.5.4 and the requirement is described in Section 5.3.

6.3.3 System Evolution Principles

Four additional principles are also presented. These principles, which are very high-level, are not specific for cellular systems, but apply to all long-lived distributed systems. These systems will undergo continuous development and the initial design had better take this into account.

System Evolution and Management

1. *Design for System Evolution*
   Cellular systems typically have an operational lifetime exceeding 20 years. During those years the system will inevitably change and so the requirements on the security protocols will change. The security protocols must therefore be designed to allow new authentication methods, new protections methods (new algorithms, etc.) and sometimes even new services (non-repudiation, etc.). To the extent possible the security protocol machinery must permit this, yet still remain secure and efficient. An important aspect of the lifetime considerations is how the security architecture handles deprecation of obsolete protocols, broken cryptographic primitives and changes to the system architecture.

2. *Backwards Compatibility*
   Backwards compatibility with old security solutions should be avoided if at all possible. It is very difficult indeed to design secure and flexible fallback schemes for use in large-scale evolved systems. The difficulty is not a theoretical one *per se*, but has to do with managing large, complex and constantly evolving networks. The difficulties come in many flavors, including the management of cut-off dates and deployment issues. Compatibility with roaming partner and the existing customer base also affects transitions between security schemes. This means that any "early" (as in weaker) security scheme will have to remain available in the network many years after it has officially been deprecated. This introduces complexity in the network and invariably limits the future design option space. Over time the added complexity can (and usually

will) lead to a system that is (partially) insecure, difficult to maintain and complex to operate.

3. *Compromise Impact Control*

   Compromise will occur. The impact must be contained. The UE population is highly distributed and exposed. Statistically some of the UEs will be compromised. The system must therefore be designed to reduce the impact of UE compromise. This would involve designs that includes perfect forward secrecy and compartmentalizing principles. Other nodes may also be compromised and the access points are vulnerable. The AN/SN operator should prepare for AP compromise events.

4. *Never Willfully Design "Weak" Versions of Algorithms or Protocols*

   Security is a weakest-link game and one should therefore never wilfully introduce weaknesses in the architecture. In particular, one must avoid the temptation to introduce weak versions of algorithms and/or security protocols. The decision to design a weak version of the A5 algorithm (A5/2) may have seemed like a good idea at the time in order to reach the export controlled markets in Eastern Europe. However, the A5/2 algorithm proved to be even weaker than the design target and it has become a liability for the GSM security architecture. The "Early-IMS Security" (EIS) solution outlined in 3GPP TR 33.978 [144] is another example where the choice to provide a weak version of the security procedure for IMS access inevitably defines the "floor" and so (due to **E.2**) it becomes harder to subsequently provide the full security solution for IMS.

The system evolution principles will be referred to as **E.X**, where X is the principle number. For example **E.4** refers to the *Never Willfully Design "Weak" Versions of Algorithms or Protocols* principle.

6.4 Summary

6.4.1 The True Value of Guidelines

In the paper "Limitations on Design Principles for Public Key Protocols" [145] Syverson warns against security guidelines and design principles. The paper discusses weaknesses of the guidelines/principles approach and provides examples where the guidelines/principles are inconsistent and incomplete.

Then what is the true value of security protocol design guidelines? Well, one should be cautious not to overplay the importance of these guidelines since they cannot be definitive. They have no formal standing and they cannot provide any guarantee that a protocol adhering to the guidelines are secure and safe. Furthermore, protocols that do break some of the guidelines may well be safe. However, as noted by Syverson [145], a set of guidelines may still have value. Syverson himself proposes a *design principles* principle:

> Use design principles at the beginning, middle, and end of designing a protocol. First, use them to guide your preliminary design. Then, when you have a specification, go through them all and look at the motivation for applying the principle in the given context. Is the motivation best served by following the principle, and if not, how might it better be served? When you have a final design, go through the principles again and look at those you violated. Make sure you have a good reason for doing so in each case. (From section 7 in "Limitations on Design Principles for Public Key Protocols" [145])

The approach suggested by Syverson is a very reasonable one. Thus, the derived guidelines are primarily seen as a design tool to aid structuring and clarifying the security protocol design.

6.4.2 Conclusion

In this chapter a set of design guidelines has been presented. The guidelines are partly generic design guidelines for security protocols and partly guidelines aimed at cellular access security protocols. The guidelines will be used as a security protocol design tool according to the Syverson proposal on how to use design guidelines.

7

Authentication and Key Agreement

Piglet sidled up to Pooh from behind. "Pooh," he whispered.
"Yes, Piglet?"
"Nothing," said Piglet, taking Pooh's paw, "I just wanted to be
sure of you."

– A.A. Milne, When We Were Very Young [146]

7.1 Introduction

The purpose of this chapter is to present the *Privacy Enhanced 3-Way Authentication and Key Agreement (PE3WAKA)* protocols. The subject of authentication and key agreement permeates Chapter 4, but conspicuously enough little is said about what exactly the PE3WAKA protocols should achieve. That is, the goals of the PE3WAKA protocols are captured indirectly and together with the requirements emerging from Chapter 5, the essence of the requirements is distilled into the set of guidelines and principles outlined in Chapter 6.

7.2 Authentication and Key Agreement

7.2.1 Background

There are numerous papers in the literature concerning authentication and key agreement/key distribution. Amongst the papers is the paper "Using Encryption for Authentication in Large Networks of Computers" by Needham and Schroeder [147] which perhaps is most famous for having (inadvertently) demonstrated that even simple toy protocols can easily be faulty.[1] The Needham-Schroeder paper did, however, inspire a lot of later papers on

---

[1] See papers by Lowe [148, 149].

the subject including the papers/reports [66, 125, 139–141, 150–153], which tackle various aspects of authentication theory including reasoning about correctness, efficiency, etc. The book *Protocols for Authentication and Key Establishment* by Boyd and Mathuria [121] is a recent and fairly comprehensive source of references on this subject.

The authentication protocol proposals can be classified as either based on public-key cryptography or on shared key symmetric cryptographic primitives. Quite a few of the proposals work under the assumption of a common Key Distribution Center (KDC). The term Authentication Centre (AuC/AC) is often used synonymously with KDC.

The purpose of corroborating the identity of a communications partner is often more a means than a goal in itself. The primary goal is frequently to set up a secure communications context that would then subsequently be used to carry out protected interaction between the parties. One often distinguishes the authentication and key distribution services as beefing either online or offline. The cellular access security services requires online authentication and key agreement.[2] Typically, one deals with key sets instead of a single key. Independent keys are then used for data integrity and for data confidentiality. One may also need to provide directional keys and there may be a need to have separate keys for control plane and user plane data. The keys in question may consist of more than the keys themselves. Commonly the cryptographic keys are encapsulated in a container object with additional data to indicate the intended usage of the keys, permitted algorithms, transfer direction, the validity of the keys, etc.

The IPsec Security Association (SA) [75, 138] is an example of a container for symmetric key protection. Correspondingly, public-key keys are commonly issued in "digital certificate" packages. The ITU-T X.509 v.3 [154] digital certificate is a typical certificate format.

7.2.2 International Standards and Recommendations

In addition to the existing cellular standards there also exist other standards for entity authentication and key agreement. The following international standards from ISO/IEC are of particular interest:

- **ISO/IEC 9798-1** [155]

---

[2] The 2G and 3G AKA protocols permits the HLR to be offline with respect to the challenge-response sequence. The PE3WAKA protocols require the HE to be online for the establishment of the medium-term security context.

Information technology – Security techniques – Entity authentication –
Part 1: General
- **ISO/IEC 9798-2** [156]
  Information technology – Security techniques – Entity authentication –
  Part 2: Mechanisms using symmetric encipherment algorithms
- **ISO/IEC 9798-3** [157]
  Information technology – Security techniques – Entity authentication –
  Part 3: Mechanisms using digital signature techniques
- **ISO/IEC 9798-4** [158]
  Information technology – Security techniques – Entity authentication –
  Part 4: Mechanisms using a cryptographic check function

Techniques similar to that of ISO/IEC 9798-2, 9798-3 and 9798-4 are deemed
suitable for use with the PE3WAKA protocols.

7.2.3 Defining the Authentication Goals

Clarifying the Authentication Goals

In their book [121, sections 2.1–2.4] Boyd and Mathuria discuss the goals of
authentication protocols, and they note that frequently the protocol designers,
the security analysts, the security protocol implementers and the security protocol users have different ideas and opinions about what exactly the protocol
achieves and, equally important, what it does not achieve. For instance, the
authors note that even the obvious goal "to authenticate" contains substantial
room for ambiguity.

Historically the goal "authenticate" has implied that keys should be distributed or generated during the authentication process. This would resonate
well with the goals for the PE3WAKA protocols, but even this type of goal is
ambiguous since both the authentication procedure itself and the key agreement can be done in different ways and with different assurance levels, etc.
For instance, there are substantial differences between online authentication
schemes vs. offline authentication schemes. The online schemes may provide
assurance of aliveness/presence of a principal entity and may guarantee that
the key material is fresh. Offline protocols cannot prove aliveness/presence
assurance and they also suffers from issues with key revocation. One must
therefore ensure that the authentication goals are clearly and unambiguously

defined. This will be helpful in the design process and it is also a prerequisite both for verification and validation purposes.[3]

Basic Authentication Goals

The fundamental goal of an AKA protocol is to perform entity authentication and to establish a security context. We shall assume that the authentication is meant to be mutual. The derived security context is bound to the verified entities, and for the context to be meaningful it must contain ways to maintain the identity corroboration. This may be achieved by subsequent use of key material bound to the entity authentication event.

In [151] Bellare and Rogaway put emphasis on the key distribution part.

> ... entity authentication is rarely useful in the absence of an associated key distribution, while key distribution, all by itself, is not only useful, but it is not appreciably more so when an entity authentication occurs along side.

The assertion that key distribution is not appreciably more useful with entity authentication is debatable, and it certainly does not fit the cellular access security model in which proving UE presence within a certain SN area is in itself an independent goal to the principals (see principle **G.4**). The Bellare–Rogaway paper also serves to point out a significant difference between the three-party protocols in the literature and the three-party case dealt with in the cellular access security context. From [151] we have:

> In this paper we realize this information advantage by way of a trusted *Key Distribution Center*, S, with whom each party P_i shares a *long-lived key*, K_i. Because of the involvement of the disinterested party S, this style of session-key distribution is called *three-party key distribution*.

As indicated above, the common assumption in the literature is that the KDC does not have an interest in the communication between the other parties *per se* – the KDC is a disinterested party that merely provides an entity authentication and key distribution service. The HE is obviously not a disinterested party, and although it will not be engaged directly in communication over the A-interface (the CA channel) the HE will certainly want a measure of control. This is explicitly captured in the *Home Control* principle (**G.1**).

---

[3] It is important to clearly distinguish "verification" and "validation" here. Verification is doing the thing right, while validation is about doing the right thing.

In [121] the authors have divided the basic authentication goals into *user-oriented goals* and *key-oriented goals*. The user-oriented goals (entity authentication goals) would concern goals like "A knows that B knows that A is the other party" and "A knows that B knows that A is the other part and that A is currently present". For the PE3WAKA protocols there are requirements both on 3-party entity authentication (and mutual knowledge thereof) and on the parties being online (freshness).

The key-oriented goals center around demonstrating that the keys are secure and fresh. The "secure" part would encompass assurance that the keys are secret (when they need to be), that the keys are authentic (key integrity assurance, etc.) and demonstration that the corresponding party has key possession. Note that for *key agreement*[4] protocols demonstration of key possession will effectively demonstrate both the "A knows that B knows . . . " and the freshness property. The same is not necessarily the case for *key distribution* protocols.

Extended Authentication Goals

According to Boyd and Mathuria [121] the basic authentication and key distribution goals simply assures that the session keys are delivered to the communication partners. Goals associated with properties of the keys (confirmed keys, freshness) are categorized as extended authentication goals.

Another set of extended goals are the so-called *responsibility* and *credit* goals. The goals *responsibility* and *credit* was originally defined by Abadi in [159]. The *responsibility* goal is (weakly) related to the non-repudiation security service, but is assumed to be met without the assistance of a third party. That is, the derived belief in *responsibility* is not dependent on any external party.

The *credit* goal is different and relates to beliefs about the message. If principal A believes that principal B transferred message M then A will give B credit for message M. This does not necessarily imply that A has authenticated B or that M contains message origin authentication implying B as the originator. It simply means that A has reason to believe that the originator was B or that M was made available to A at Bs request. As noted by Abadi the notions of *responsibility* and *credit* are easiest to distinguish for

---

[4] Understood here to denote the case that the key is not unilaterally decided by any one party, i.e., the demonstration of key possession can be interpreted as proof that the keys are fresh.

public-key systems.[5] A recent account of the *responsibility* and *credit* goals is found in [160]. The *responsibility* and *credit* goals are not seen as relevant goals for the PE3WAKA protocols.

Robustness and Resilience Goals

In section 2.4 of [121] the authors discuss goals related to key compromise. In a larger perspective the goals mentioned in the same section of [121] are really about robustness and resilience of the protocol in the face of a capable intruder that will exploit any shortcomings or failures of the system. In large distributed systems there will inevitably be failures and compromise. The failure modes are many and can originate from everything from incorrect specification to incorrect implementation. Other failure modes include incorrect assumptions, etc. For instance, it is quite common to assume that all principals are honest (this assumption underlies the BAN logic [66]), but this is not likely a correct trust assumption in a system with many millions of principal entities. Another aspect of the sheer volume of subscribers is that some of the devices (MT and/or UE) are almost certain to become physically compromised in some way or another. One therefore cannot assume that all principals will act "honestly".

The PE3WAKA protocols should therefore be designed to be robust and resilient in face of compromised/dishonest principals. To this end the PE3WAKA protocol must ensure that the compromise of one protocol run does not unduly affect future and/or previous protocol runs. Likewise, the impact from a compromised cryptographic algorithm should not (unduly) prevent the protocol from succeeding with another algorithm.

The weakest part in the system is likely to be the UE/SM and with the number of UEs accommodated it is crucial that the compromise of one UE/SM does not affect other UEs. For a system that must undergo cycles of evolution and development the need to address issues like introduction of new algorithms (and deprecation of old algorithms) is urgent.

Resilience to Denial-of-Service (DoS) Attacks

Another aspect is resilience against DoS attacks. The PE3WAKA protocols must be constructed to mitigate the effect of DoS attacks. There are several types of DoS attacks; including attacks on computational capacity and attacks on communications capacity.

---

[5] "The two facets of authentication are most clearly separate in protocols that rely on asymmetric cryptosystems, such as the RSA cryptosystem" [159].

With respect to defense against attacks on computational capacity it is useful if the PE3WAKA protocols could reduce the instantaneous computational requirements. Amongst the techniques that may be helpful here is to design the protocol such that the principal entities may pre-compute certain values (DH parameters, etc.) and to postpone certain requests (rekeying), etc. It is also useful to balance the PE3WAKA protocols such that the attacker must maintain state and/or itself be forced to carry out a substantial amount of work (like executing cryptographic primitives).

The communication capacity would only be an issue locally in the AN. That is, it would seem unlikely that an attacker would be able to consume the communication capacity with PE3WAKA requests any other place than over the radio link. The radio link is vulnerable to this type of attack, but the communications capacity requirements of the PE3WAKA requests are modest even when viewed from the AN point of view. It may be rightly argued that the "setup" channel capacity is limited and that it is certainly possible to overwhelm the capacity of this channel locally. To attack the SN and HE as such the intruder would have to, in a geographically distributed manner, attack a large set of setup channels.[6]

It is noted that if service disruption alone is the goal then the intruder could achieve this just as effectively with less sophisticated attacks. On an SN system level the best practical protection against DoS attacks would probably be to limit the number of PE3WAKA request per time unit per AP. From the HE point of view one may limit the number of PE3WAKA requests from different SNs per time unit.

Efficiency Goals

It is imperative that the PE3WAKA protocols be efficient. Efficiency and performance concerns were also discussed and addressed in Section 4.8. A particular concern for the cellular environment is the context setup delays. The setup delay is dependent on several factors, of which the number of exchanged messages (round trips) is a significant factor. In [152] Gong provides lower bounds on the number round trips necessary to complete the protocols. Gong distinguishes the protocols according to the goals they have and it is no surprise that the lower bounds vary accordingly. Gong classifies protocols [152] according to the characteristics given in Figure 7.1.

---

[6] One might term such an attack a Geographically Distributed Denial-of-Service (GDDoS) attack.

- **NB** – Nonce Based
- **TB** – Time Based (clock based)
- **AH** – Authentication + Handshake
- **AO** – Authentication Only
- **SO** – Server Only (key chooser)
- **CO** – Client Only (key chooser)
- **CC** – Both clients (key chooser)

Figure 7.1 Categories in the Gong classification scheme.

The PE3WAKA protocols do not match with the cases identified by Gong, but the closest case would the one discussed in "Case 12: NB+AH+CC" in [152, section 3.12]. There Gong finds that at least six messages and that at least five rounds is necessary to complete a "NB+AH+CC" protocol. The PE3WAKA protocol also has extended goals not accounted for in the cases listed by Gong in that the PE3WAKA protocols must provide subscriber privacy and subscriber identity management. One should therefore expect the PE3WAKA protocols to require additional messages and round-trips.

Another point worth mentioning is that for a real-world case one should expect to see different propagation delay factors for the A-interface (CA channel) and the B-interface (CB channel), and so some messages may be more costly to execute than others based on the channel they pass over.

Privacy Goals

The privacy goals for the PE3WAKA protocol is discussed in Chapter 5 and explicitly expressed in Chapter 6 (Guidelines **G.14–G.18**). In line with the privacy guidelines for the PE3WAKA protocols the following is noted:

(a) The $MTC_{(HE,SN,UE)}$ context shall not contain UE location information or the permanent UE identity.
(b) The $STC_{(SN,UE)}$ context shall not contain the permanent UE identity. The $STC_{(SN,UE)}$ *may* contain UE location information (i.e. related to expiry condition for the context).

7.2.4 The Goals for the PE3WAKA Protocols

The primary goal of the PE3WAKA protocols is to establish the medium-term security context between the three cellular system principal entities. The larger context for the cellular systems means that there are legal and regulatory requirements on the systems. These include extended support for emergency services and for lawful interception, etc. It also includes provisions to protect the customer with respect to the network operators. One aspect of this is the provision to protect the privacy of the subscribers. The scheme selected for protection of the subscriber identity is derived from the "random reference identity" outline in Section 5.4.2.

The goal of the PE3WAKA protocols is captured and summed up in the guidelines presented in chapter 6. It is noted that the PE3WAKA goals, as captured in the guidelines, need not all be fulfilled for all instances of the protocol. Indeed, as Syverson indicated in [145] (quoted in Section 6.4), it may be permissible to break a principle provided one knows why it was done and provided it can be explicitly justified.

7.2.5 Secrecy Classification

In the following the security service "data confidentiality" is classified into four different sub-categories:

- *Secret*,
- *Subscriber Privacy*,
- *Non-Public*,
- *Public*.

The *secret* category is the classical data confidentiality requirement along the lines of "The property that information is not made available or disclosed to unauthorized individuals, entities, or processes" [161].

The *subscriber privacy* category have the same characteristics as *secret*, but the implications of a breach would be different. For the *secret* category one should assume that the whole protocol instance, potentially also past and future instances, would be compromised if a *secret* information element (IE) is exposed. Should a *subscriber privacy* IE be exposed the implication is that one must assume that the privacy associated with the IE is broken. The protocol may still be able to provide authentication and key agreement services.

For the *non-public* category one would find information elements that the entities does not want to disclose unnecessarily. That is, the information is not secret *per se*, but one still does not want it published indiscriminately. An example of this would be information concerning security policy preferences or details of the network topology. The protection requirements and methods used for *non-public* data may differ from the *secret* and *subscriber privacy* categories. In fact, depending on circumstances, it may be permissible to transmit *non-public* data in clear. Thus, *non-public* data may not always need cryptographic protection.

The classification *public* is used for data that is intended or permitted to be fully public. That is, there is absolutely no harm done if a *public* IE is known to any given intruder.

7.3 Cryptographic Basis for the PE3WAKA Protocols

7.3.1 Effective Cryptography

As is common it is assumed that the cryptographic primitives are effective. That is, it is assumed that the cryptographic primitives themselves will not succumb to attacks that are significantly easier to mount than brute force attacks.

7.3.2 Pseudo-Random Functions

Pseudo-random functions are used extensively in the PE3WAKA protocols. The requirements upon the $prf()$ function is similar to the requirements on the 3G $f0$ function (2.8) described in Section 2.6.2. To comply with the **G.10** requirement for 128-bit cryptographic strength, the $prf()$ function must be able to produce 128-bit wide pseudo-random numbers. Equation (7.1) shows the generic $prf()$ function. Unless otherwise stated the output, rnd, is a 128-bit wide field.

$$prf(\cdot) \rightarrow rnd \qquad (7.1)$$

7.3.3 Key Derivation

A generic key derivation function, $kdf()$, for symmetric key generation is defined to be used in the PE3WAKA protocols. The $kdf()$ function shall be fully deterministic and it shall take as input a shared secret s and a tuple of arguments (arg). The output is a symmetric key key. The specific cryptographic requirements on the $kdf()$ functions are complex and shall not be further examined here, but suffice to mention that there must be no apparent correlation between the input, including the secret key, and the output. Furthermore, it must be computationally infeasible to derive s from knowledge of arg and key. This should hold true even for long sequences of (arg,key) parameters.

The base secret in the PE3WAKA protocols will be a secret generated from a Diffie–Hellman exchange, and to allow for multiple derived sessions, the DH secret is assumed to be 256 bits wide. The arguments (arg) should themselves contain at least 128 bits of entropy. In practice, this requirement is fixed to mean that the arguments should contain at least one 128-bit wide pseudo-random information element. The IE providing the entropy should be specifically designated to do so.

$$kdf_s(arg) \rightarrow key \qquad (7.2)$$

Equation (7.2) shows the generic $kdf()$ function. The output is a 128-bit secret key key.

7.3.4 Mutual Challenge-Response Schemes

The PE3WAKA protocols are 3-way protocols, but there is still a need for 2-way mutual authentication between the UE and the HE and between the SN and the HE. The PE3WAKA protocols work under the assumption of a pre-established security context between the SN and the HE.

The (UE, HE) mutual authentication sequence can be performed in several ways. However, there is no compelling reason to change from the symmetric-key challenge-response schemes used in 3G authentication to provide entity authentication between the UE and the HE. The methods are computationally inexpensive and require comparatively little overhead on the communication channel.

It would be desirable to avoid the complication of a sequence number scheme as found in the 3GPP/3GPP2 AKA protocols. Unless properly configured the sequence number schemes can trigger unnecessary synchronization events or worse, they may fail to protect against the use of aging security credentials. Use of sequence numbers is therefore avoided in the PE3WAKA protocols. Instead, a much simpler online double challenge-response scheme is used to provide mutual entity authentication with guaranteed freshness. There is a cost in terms of the need for an additional signalling pass, but this is effectively mitigated in a combined identity management, mobility management and security establishment protocol.

The mutual challenge-response sequence requires a response function $res()$ (7.3) that takes the random challenge rc and produces a signed response sr under control of a shared authentication secret as. The challenge, rc, provided to $res()$ is assumed to be a tuple and it is required that at least one of the tuple elements must be pseudo-random (i.e. produced by $prf()$). The cryptographic requirements on $res()$ is similar to that of the UMTS $f2()$

function (2.12).

$$res_{as}(rc) \rightarrow sr \tag{7.3}$$

7.3.5 Message Authentication

The following function is a generic message authentication code function. The function, under control of the shared key k, computes a cryptographic integrity check value (icv) over the content of the message msg. The service provided is primarily *message data integrity*, but is can also provide assurance on message origin and intended message destination if the address fields are included in the message. The icv is appended to the original message to form the new integrity protected message: $msg_{new} := msg||icv$. The width of the icv should be at least 64 bits (see Guideline **G.10**). It is noted that the requirements pertaining to (7.4) are otherwise similar as to those for the UMTS $f1$ function (see Section 2.6.2).

$$MAC_k(msg) \rightarrow icv \tag{7.4}$$

7.3.6 Diffie–Hellman (DH) Exchange

The Generic DH Exchange

The Diffie–Hellman exchange was first described in Diffie and Hellmans seminal 1976 paper "New Directions in Cryptography" [162]. The DH exchange is thoroughly described and analyzed in the literature. One example is the discussion found in [22]. Suffice here to say that in a basic DH exchange Alice and Bob must first agree on an appropriate prime p and a suitable generator g ($2 \leq g \leq p - 2$). Alice chooses a suitable random secret a such that $A = g^a \bmod p$. Bob chooses a suitable random secret b such that $B = g^b \bmod p$. Alice and Bob then exchange A and B, and computes the DH secret s as shown below. The basic DH exchange is not authenticated and is therefore susceptible to man-in-the-middle attacks.

$$s = B^a \bmod p$$

$$s = A^b \bmod p.$$

The medium-term security context will have a DH secret s as its basis. A standard 2-way DH exchange will be used. There are three distinct cases:

(A) Execute the DH exchange over the A-interface (the over-the-air CA-channel)

(B) Execute the DH exchange over the B-interface (the fixed CB-channel)

(C) Execute the DH exchange over the A- and B-interface (the CC-channel)

Case A is problematic. We will want to have a shared secret s with an entropy in the order of 256 bits. With standard DH methods this will amount to an exchange of at least two 15 Kbits information elements over the A-interface [163]. During setup, this may not be acceptable. However, if one uses an elliptic curve cryptography based DH method, one may be able to reduce the information element size considerably [163].

Case B is simpler. One does not have to worry about information element sizes as the bandwidth between SN and HE is assumed to be sufficient for any PE3WAKA needs. The problem with case B is that the UE needs access to the secret. So, the HE must be able to forward the secret to UE in a confidentiality and integrity protected manner.

Case C has all the problems associated with cases A and B. There is also an advantage to running the DH exchange between the UE and the HE in that these two principals already have an established long-term context. That is, the authentication of the DH exchange ought to be relatively simple to achieve (the same holds for case B).

The Diffie–Hellman Functions
The following generic DH functions are defined:

$$DHparam_{(group)}(\cdot) \rightarrow (n, N) \tag{7.5}$$

$$DHcalc_{(group)}(n, N) \rightarrow dhs \tag{7.6}$$

In (7.5) the *group* parameter identifies the DH group. The derived tuple (n, N) would then either be (a, A) for Alice or (b, B) for Bob. In (7.6) the *group* parameter identifies the DH group. The input parameters (n, N) would then either be (a) (a, B) for Alice or (b) (b, A) for Bob. The derived parameter *dhs* is the shared Diffie–Hellman secret. It is noted that pre-computation of the DH parameters is possible provided that the parties can agree upon the DH group prior to PE3WAKA execution. Pre-computation can be an effective way of balancing the load and it can also serve as a DoS protection mechanism.[7]

---

[7] In [142] the authors specifically describe the use of a FIFO queue of pre-computed DH parameters to prevent DoS attacks. The principal must be able to replenish the FIFO queue at a rate that will suffice for legitimate busy-hour usage.

Use of DH in Protocol Design

The DH exchange is an excellent method for providing shared key material between two principals. Since the basic DH exchange is not authenticated it is essential to add message integrity and message origin authentication to the basic exchange.

The DH exchange is notably used in the IPsec protocol suite. That is, it is used in the Internet Key Exchange (IKE) protocol which is part of the IPsec protocol suite. IKE version 1 is captured in RFC 2409 [137] and IKEv2 is found in RFC 4306 [138]. Both the IKE versions are based on the so-called SIGMA class of protocols. The SIGMA approach is described in detail in [130]. The SIGMA protocols draw on experience with the Station-To-Station (STS) protocols. The STS protocols and the design rationale behind the STS protocols is described in [164]. It is noted that both the STS paper [164] and SIGMA paper [130] contain detailed discussions about security protocol requirements, attacks models, etc., and these papers are therefore well worth studying.

7.3.7 Standard Symmetric-Key Encryption/Decryption

Standard symmetric-key cryptography is used in the PE3WAKA protocols and it is extensively used in the link layer protection. The PE3WAKA requirements will otherwise resemble the requirements found in the 3G specification (see 3GPP TS 33.105 [42]). In Equations (7.7) and (7.8) M is the plaintext message and C is the ciphertext message. As is the usual case with symmetric-key cryptography the secret key k is shared and common to the encryption and decryption operations. Thus, $E_k(X) = D_k(X)$.

$$E_k(M) \rightarrow C \tag{7.7}$$

$$D_k(C) \rightarrow M \tag{7.8}$$

7.3.8 Standard Public-Key Encryption/Decryption

Standard public-key cryptography is used in some of the PE3WAKA protocols. In Function 7.9 and 7.10 M is the plaintext message and C is the ciphertext message. In public-key cryptography one uses distinct keys for encryption (public key k) and decryption (private key k^{-1}), where $k \neq k^{-1}$. The use of standard public-key cryptography in the PE3WAKA protocols is limited to providing *subscriber privacy* (see Section 7.2.5).

Table 7.1 The core cryptographic functions.

Function	Func. Ref.	Comment
$prf(\cdot) \rightarrow rnd$	(7.1)	Derivation of pseudo-random number
$kdf_s(arg) \rightarrow key$	(7.2)	Key derivation function
$res_{as}(rc) \rightarrow sr$	(7.3)	Response function
$MAC_k(MSG) \rightarrow ICV$	(7.4)	Generate (cryptographic) Integrity Checksum Value
$DHparam_{(group)}(\cdot) \rightarrow (n, N)$	(7.5)	Generate DH private/public param. pair
$DHcalc_{(group)}(n, N) \rightarrow dhs$	(7.6)	Calculations of shared DH secret
$E_k(M) \rightarrow C$	(7.7)	Symmetric-key encryption
$D_k(C) \rightarrow M$	(7.8)	Symmetric-key decryption
$E_k(M) \rightarrow C$	(7.9)	Public-key encryption
$D_{k^{-1}}(C) \rightarrow M$	(7.10)	Public-key decryption

$$E_k(M) \rightarrow C \qquad (7.9)$$

$$D_{k^{-1}}(C) \rightarrow M \qquad (7.10)$$

7.3.9 Summary of the Core Cryptographic Functions

The core cryptographic functions used in the PE3WAKA protocols are listed in Table 7.1.

7.4 The PE3WAKA Protocol Family

7.4.1 Introduction

The PE3WAKA protocol family shares a set of common features, but there is also plenty of room for adaptations and modifications. The set of common features include:

- The principals: HE, SN and UE;
- The Initiator-Responder model;
- The trust assumptions, the intruder and attack models;
- The long-term security contexts ($LTC_{(HE,UE)}$ and $LTC_{(HE,SN)}$)
- The communication architecture: A-Interface and B-interface; Channels: CA, CB and CC;
- The 3-way online authenticated medium-term security context;
- The need for key derivation for the short-term contexts;
- A secured CB channel (authenticated, with confidentiality- and integrity protection);

- The efficiency requirements (round-trips), etc.;
- The subscriber privacy requirements (basic requirements);

7.4.2 A Brief Classification of the PE3WAKA Protocol Family

The following classification is not intended to be a complete taxonomy of the PE3WAKA protocol family, but is rather aimed at capturing the major differences between the various PE3WAKA protocols. The main differentiating factors are briefly discussed in the following.

Derivation of Context Secret

The $MTC_{(HE,SN,UE)}$ context is to be used to derive short-term contexts. To this end a shared secret is used as the context basis.

For the PE3WAKA protocols the choice has been to use DH methods to derive the shared secrets. The DH exchange has many attractive cryptographic properties and the theory for DH methods is well established. In principle one could have used a tripartite DH method, but as was argued in Section 4.7.2 the tripartite DH protocol model does not fit well with the PE3WAKA requirements. However, one may still use a traditional DH exchange between two of the principals and then let the third principal learn the secret from one of the other principals. It is also noted that the $MTC_{(HE,SN,UE)}$ context is primarily used to derive $STC_{(SN,UE)}$ contexts. That is, it may not be strictly necessary for the HE to know the DH secret. For the PE3WAKA protocols there are three distinct cases:

DH-secret negotiation:

- Over the CA channel;
- Over the CB channel;
- Over the CC channel.

Authentication Method and Signature Method

The CB channel is assumed to be established and protected prior to running PE3WAKA. The mutual entity authentication procedure between the UE and the HE is based on a challenge-response mechanism. There are several choices as how to computing the response, including using encryption methods, using message authentication (MAC) methods and using public-key methods.

Challenge-response; computing the response:

- *Encryption*
  Some simple operation is performed on the challenge and then the result is encrypted, using a symmetric encryption primitive under control of a secret key, before being returned. This scheme would be in line with ISO/IEC 9798-2 [156].
- *Public-key Signing*
  A public-key signature is computed over the challenge. The method is computationally expensive and may generate relatively large checksums. This scheme would be in line with ISO/IEC 9798-3 [157].
- *Message Authentication*
  The response is computed over the challenge under control of a secret key using a MAC function. This scheme would be in line with ISO/IEC 9798-4 [158].

PE3WAKA Classification Overview

Table 7.2 presents an overview of the basic PE3WAKA protocols. The implications of changing the parties in the DH exchange are quite pronounced. If fact, the execution logic of the protocol must be entirely redesigned since the use and appropriate binding of the derived DH secret is fundamental to the medium-term context. The impact on changing response method is comparatively small.

Table 7.2 The PE3WAKA protocols.

Protocol	DH exchange	Response
A1	UE–SN	Encryption
A2	UE–SN	MAC signed
A3	UE–SN	Public-key Signed
B1	SN–HE	Encryption
B2	SN–HE	MAC signed
B3	SN–HE	Public-key Signed
C1	UE–HE	Encryption
C2	UE–HE	MAC signed
C3	UE–HE	Public-key Signed

7.4.3 Engineering Decision

The following engineering decisions may not be strictly necessary, but they certainly are in line with the *prudent engineering* principles outlined in [139]. The principles, in an abridged form, are also presented in Section 6.2.

One decision is to have explicit identity binding. This decision is reflected in **PE.1**. The binding of identities is done multiple times. This may seem excessive, but it is noted that (a) there is virtually no additional cost to include the identities in the entity authentication and key derivation functions and that (b) from an engineering viewpoint it makes sense to express the essential bindings explicitly. This is in also line with **PE.3**.

The final set of considerations concerns the use of randomness and uniqueness. There are several identifiers that are derived by a pseudo-random function. Still, the intended use of the uniqueness/randomness characteristics in the different element is not the same. Consequently, it has been a conscious engineering decision to use separate information element when a combined element may have sufficed. This is in line with **PE.1** and **PE.6**.

7.4.4 Example Protocol

The PE3WAKA Example Protocol

In the following section we will go through a detailed discussion of one Privacy Enhanced Authentication and Key Agreement (PE3WAKA) protocol. The example is merely to show one possible way to construct a privacy enhanced AKA protocol; it is not *the way* to construct a PE3WAKA protocol. The example protocol may appear to be quite complete, but it is in fact missing a number of features that must be incorporated in a real-world protocol. Amongst the missing parts are elements such as protocol identifier, protocol version identifier, transaction numbering, etc. Furthermore, for a real protocol one would obviously have to standardize IE type systems and IE encoding schemes, and issues like the endian choice, etc. A robust real-world protocol must also have fully defined exception handling.

Protocol Presentation: Symbols and Notation

The following notations and symbols are used in the protocol description and discussion. Additionally, the so-called Alice-Bob notation is used. This includes the convention to denote encrypted content by curly brackets ({ and }) and to use subscripts to indicate use of cryptographic keys. Subscripts are sometimes also used to indicate which entity created the information element. It is sometimes useful to distinguish a received value from a locally calculated one. This is done by prepending an X (eXpected) to the name of the locally calculated variable. That is, if the received variable is denoted RES then the corresponding locally derived variable would be denoted $XRES$.

|	"or". Used as "choice" operator.
||	String concatenation
$\wedge$	"and-then" concatenation of expressions
:=	Assignment (to the left)
==	Comparison
a, b	comma separated n-tuple
$\rightarrow$	(1) Messages: denotes sending from/to.
	(2) Functions: function output to the righthand side.

7.5 Protocol PE3WAKA_B2

7.5.1 High-Level Description

The following protocol is based on standard public-key methods and uses a MAC signed challenge-response scheme. The DH exchange is executed over the CB-channel between the SN and the HE. This means that the DH secret must be forwarded to the UE from the HE in an authenticated and secure manner.

Outline in Augmented Alice-Bob Notation

Figure 7.2 presents the PE3WAKA_B2 protocol in the Alice–Bob notation and Figure 7.3 gives a graphical depiction of the protocol. The identifiers are described in Table 7.3.

> (M1) UE $\rightarrow$ SN: M1($HEID, A, VP$)
>   Where: $A := \{UEID, CID, CH_{UE}, ICV_{M1}\}_{HEK}$
> (M2) SN $\rightarrow$ HE: $\{$M2(TI_{CB}, A, VP, DH_{SN})$\}_{bkey}$
> (M3) HE $\rightarrow$ SN: $\{$M3(TI_{CB}, CID, B, DH_{HE})$\}_{bkey}$
>   Where: $B := \{RES_{UE}, CH_{HE}, dhs, CID\}_{cck_{HE}}$
> (M4) SN $\rightarrow$ UE: M4($B, \{CID, SID\}_{cak_{SN}}$)
> (M5) UE $\rightarrow$ SN: $\{$M5($SID, \{RES_{HE}, CID\}_{cck_{UE}}$)$\}_{cak_{UE}}$
> (M6) SN $\rightarrow$ HE: $\{$M6($CID, \{RES_{HE}, CID\}_{cck_{UE}}$)$\}_{bkey}$
> (M7) HE $\rightarrow$ SN: $\{$M7(CID, ICV_{M7})$\}_{bkey}$
> (Mn) SN $\rightarrow$ UE: $\{$Msg($MSG\_CONTENTS$), $ICV_{M7}\}_{sk_{SN}}$

Figure 7.2 The PE3WAKA_B2 protocol in Alice–Bob notation.

Table 7.3 PE3WAKA_B2 information elements.

Identifiers	Comment/Definition
$HEID, SNID$	Public HE identity; Public SN identity;
$UEID$	Private UE identity (only known to UE and HE);
CID	Context Identity (representing the UE); Associated with the $MTC_{(HE,SN,UE)}$ context;
SID	Session Identity (representing the UE); Associated with the $STC_{(SN,UE)}$ context.
CH_{UE}, RES_{UE}	Random challenge, generated by UE, and the corresponding response;
CH_{HE}, RES_{HE}	Random challenge, generated by HE, and the corresponding response;
$autk$	UE authentication key; part of $LTC_{(HE,UE)}$ context
$bkey$	Key associated with B-interface; Protecting the CB channel;
cak_{UE}, cak_{SN}	Directional shared key for CA-channel; Derived for $MTC_{(HE,SN,UE)}$ context;
cck_{UE}, cck_{HE}	Directional shared key for CC-channel; Derived for $MTC_{(HE,SN,UE)}$ context;
sk_{UE}, sk_{SN}	Directional session keys; Short-term context ($STC_{(SN,UE)}$) keys;
dh_X, DH_X	Private and Public DH parameters generated by entity X;
dhs	DH secret; Generated between HE and SN;
HEK, HEK^{-1}	HE public-key key pair
AC	Area code; Defined for an AN area.
VP_{HE}, VP_{SN}	Validity Period value for HE and SN respectively; ($VP := Min(VP_{HE}, VP_{SN})$;
A, B	Encrypted blocks containing several IEs
TI_{CA}, TI_{CB}	Transaction Identifier for CA and CB channel.
ICV_{Mx}	Integrity Check Value; for message "Mx";
DIR	Direction indication: (UE→HE), (HE→UE), (UE→SN) or (SN→UE);
$BIND$	$BIND := HEID, SNID, CID, VP$

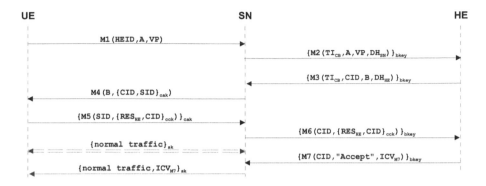

Figure 7.3 Outline of the PE3WAKA_B2 protocol.

7.5.2 Detailed Description

The following is a detailed description of the PE3WAKA_B2 protocol.

Information Elements

The information elements used in the PE3WAKA_B2 protocol are presented in Table 7.3.

Initial Entity Knowledge: The Long-Term Context

This is a summary of the prerequisite knowledge/state prior to executing the PE3WAKA_B2 protocol (Table 7.4). Parameters concerning DH groups, etc., are not addressed below, but in practice all three entities must know the DH group parameters. The DH group parameters are considered to be "Non-Public". UE knowledge is stored in SM. The SN also knows the AP area codes (AC) and other data related to the access network (AN). AN network configuration information is considered public knowledge.

Cryptographic Functions

The cryptographic functions used in the protocol are listed in Table 7.5.

Step-by-Step

1. *Message M1*
   UE constructs block A;

   – Read from SN broadcast channel: $SNID$ and VP_{SN};
   – $f0(\cdot) \rightarrow CID$;

Table 7.4 Initial knowledge.

Entity	Data	Context	Security Requirement
UE	$UEID$	$LTC_{(HE,UE)}$	Subscriber Privacy;
UE	$HEID$	$LTC_{(HE,UE)}$	Public information;
UE	VP_{HE}	$LTC_{(HE,UE)}$	Non-Public;
UE	sp	$LTC_{(HE,UE)}$	Non-Public;
UE	HEK	$LTC_{(HE,UE)}$	Non-Public;
UE	$autk$	$LTC_{(HE,UE)}$	Secret;
SN	$SNID, HEID$	$LTC_{(HE,SN)}$	Public information;
SN	VP_{SN}	SN	Public information;
SN	$bkey$	$LTC_{(HE,UE)}$	Secret;
HE	$UEID$	$LTC_{(HE,UE)}$	Subscriber Privacy;
HE	$HEID$	$LTC_{(HE,UE)}$	Public information;
HE	VP_{HE}	$LTC_{(HE,UE)}$	Non-Public;
HE	s	$LTC_{(HE,UE)}$	Secret;
HE	sp	$LTC_{(HE,UE)}$	Non-Public;
HE	$autk$	$LTC_{(HE,UE)}$	Secret;
HE	HEK	$LTC_{(HE,UE)}$	Non-Public;
HE	HEK^{-1}	$LTC_{(HE,UE)}$	Secret;
HE	$SNID, HEID$	$LTC_{(HE,SN)}$	Public information;
HE	$bkey$	$LTC_{(HE,SN)}$	Secret;

- $f1(\cdot) \rightarrow CH_{UE}$
- $VP := Min(VP_{SN}, VP_{HE}) \wedge$
  $BIND := HEID, SNID, CID, VP;$
- $f3_{autk}(BIND, UEID, (UE \rightarrow HE)) \rightarrow cck_{UE};$
- $f3_{autk}(BIND, UEID, (HE \rightarrow UE)) \rightarrow cck_{HE};$
- $MAC_{cck_{UE}}(BIND, UEID, CH_{UE}) \rightarrow ICV_{M1};$
- $E_{HEK}(UEID, CID, CH_{UE}, ICV_{M1}) \rightarrow A;$

Table 7.5 PE3WAKA_B2 functions.

Function/Definition	Comment	Func.ref
$f0: prf(\cdot) \rightarrow CID\|SID\|TI_{CB}$	Construction of pseudo-random IEs	(7.1)
$f1: prf(\cdot) \rightarrow CH_{UE}\|CH_{HE}$	Construction of challenge data;	(7.1)
$f2: kdf_{dhs}(BIND, TI_{CA}, DIR) \rightarrow cak$	Derivation of shared key between UE and SN;	(7.2)
$f3: kdf_{autk}(BIND, UEID, DIR) \rightarrow cck$	Derivation of shared key between UE and HE;	(7.2)
$f4: kdf_{dhs}(SID, CID, AC, DIR) \rightarrow sk$	Derivation of session key between UE and SN;	(7.2)
$f5: res_{autk}(BIND, UEID, CH_X) \rightarrow RES_X$	Response to challenge (X=UE or X=HE);	(7.3)
$f6: DHparam(\cdot) \rightarrow (dh_N, DH_N)$	Generated by $N \in$ (SN,HE);	(7.5)
$f7: DHcalc(dh_N, DH_{N'}) \rightarrow dhs$	Calculated by SN and HE;	(7.6)
$MAC: MAC_{key}(message) \rightarrow ICV$	Integrity check value for *message*;	(7.4)
$E: E_{key}(M) \rightarrow C$	Symmetric-key encryption	(7.7)
$D: D_{key}(C) \rightarrow M$	Symmetric-key decryption	(7.8)
$E: Encrypt_{HEK}(UEID, CID, CH_{UE}, ICV) \rightarrow A$	Used by UE to protect $UEID$, etc.;	(7.9)
$D: Decrypt_{HEK^{-1}}(A) \rightarrow UEID, CID, CH_{UE}, ICV$	Used by HE to decrypt A;	(7.10)

Observe that the UE unilaterally decides the *cck* key set. A transaction identifier, TI_{CA}, is constructed from the cell identifier which includes the area code AC, the physical radio channel number and further qualified by the logical channel identifiers frame number, etc. This identifier is not transferred explicitly, but will be constructed and used by the UE and the SN in local key derivation. The HE will not normally know TI_{CA}.

THEN: UE $\rightarrow$ SN: M1($HEID, A, VP$)

2. *Message M2*
   SN *sees* the claimed IEs A, $HEID$ and VP. SN develops no new beliefs.
   SN verifies that SN has roaming agreement with the HE and that the proposed validity period is acceptable ($VP \leq VP_{SN}$). SN constructs the transaction identifier, TI_{CA} with data received from the AN. SN allocates a unique transaction identifier, TI_{CB}, for use over the CB channel. TI_{CB} is only used until the SN learns the CID from the HE.
   SN constructs DH parameters:

       – $f6(\cdot) \rightarrow dh_{SN}, DH_{SN}$

   THEN: SN $\rightarrow$ HE: $\{$M2(TI_{CB}, A, VP, DH_{SN})$\}_{bkey}$

3. *Message M3*
   HE decrypts the outer message and *sees* message **M2**. HE knows that SN *once said* the contents of **M2**. HE creates an authentication request context and temporarily assigns TI_{CB} to the context. HE proceeds to decrypt and verify A:

       – Decrypt A: $D_{HEK^{-1}}(A) \rightarrow UEID, CID, CH_{UE}, ICV_{M1}$

   Assume $UEID$ to be valid: retrieve long-term context associated with $UEID$. Then compute keys and verify ICV_{M1}.

       – $BIND := HEID, SNID, CID, VP$;
       – $f3_{autk}(BIND, UEID, (\text{UE}\rightarrow\text{HE})) \rightarrow cck_{UE}$;
       – $f3_{autk}(BIND, UEID, (\text{HE}\rightarrow\text{UE})) \rightarrow cck_{HE}$;
       – $MAC_{cck_{UE}}(BIND, UEID, CH_{UE}) \rightarrow XICV_{M1}$;
       – Verify: $ICV_{M1} == XICV_{M1}$;

At this stage HE knows that the contents of A were uttered by UE. The HE therefore tentatively accepts and assigns CID as an alias identity for $UEID$. It generates a reply challenge and a response to the given challenge.

- $f1(\cdot) \rightarrow CH_{HE}$;
- $f5_{autk}(BIND, UEID, CH_{UE}) \rightarrow RES_{UE}$;
- $f5_{autk}(BIND, UEID, CH_{HE}) \rightarrow RES_{HE}$;

The HE then computes the DH secret and constructs block B.

- $f6(\cdot) \rightarrow dh_{HE}, DH_{HE}; \wedge\ f7(dh_{HE}, DH_{SN}) \rightarrow dhs$;
- $E_{cck_{HE}}(RES_{UE}, CH_{HE}, dhs, CID) \rightarrow B$;

THEN: HE $\rightarrow$ SN: $\{\mathrm{M3}(TI_{CB}, CID, B, DH_{HE})\}_{bkey}$

4. *Message M4*
   SN decrypts the outer message and *sees* the contexts of **M3**. SN therefore knows that HE uttered **M3**. Due to the inclusion of TI_{CB} the SN also knows that the information is fresh and applies to the current protocol run. The SN tentatively accepts CID and proceeds to compute the DH secret:

   - $f7(dh_{SN}, DH_{HE}) \rightarrow dhs$;

   SN then constructs the session identity, SID, and the *cak* and *sk* key sets:

   - $f0(\cdot) \rightarrow SID$;
   - $BIND := HEID, SNID, CID, VP$;
   - $f2_{dhs}(BIND, TI_{CA}, (\mathrm{SN}\rightarrow\mathrm{UE})) \rightarrow cak_{SN}$
   - $f2_{dhs}(BIND, TI_{CA}, (\mathrm{UE}\rightarrow\mathrm{SN})) \rightarrow cak_{UE}$
   - $f4_{dhs}(SID, CID, AC, (\mathrm{SN}\rightarrow\mathrm{UE})) \rightarrow sk_{SN}$
   - $f4_{dhs}(SID, CID, AC, (\mathrm{UE}\rightarrow\mathrm{SN})) \rightarrow sk_{UE}$

   THEN: SN $\rightarrow$ UE: $\mathrm{M4}(B, \{CID, SID\}_{cak_{SN}})$

5. *Message M5*
   UE receives **M4** and sees the (encrypted) contents. The UE then processes block B.

   - $D_{cck_{HE}}(B) \rightarrow RES_{UE}, CH_{HE}, dhs, CID$

   The presence of CID implies that the HE has successfully decrypted block A. It also means that the HE has tentatively accepted CID. UE verifies the given response and computes a response itself. Subsequent to this the UE will consider the HE to be authenticated.

- $f5_{autk}(BIND, UEID, CH_{UE}) \rightarrow XRES_{UE} \land$ Verify: $XRES_{UE} == RES_{UE}$
- $f5_{autk}(BIND, UEID, CH_{HE}) \rightarrow RES_{HE}$

The UE proceeds to derive *cak* key sets.

- $f2_{dhs}(BIND, TI_{CA}, (SN \rightarrow UE)) \rightarrow cak_{SN}$
- $f2_{dhs}(BIND, TI_{CA}, (UE \rightarrow SN)) \rightarrow cak_{UE}$

The UE now decrypts the contents of the SN encrypted block.

- $D_{cak_{SN}}(CID, SID) \rightarrow CID, SID$

The UE can then infer that the SN must have gotten CID from the HE and that SN have access to *dhs*. The UE therefore assumes that SN was trusted by HE. The UE knows that CID is fresh and it therefore assumes the SN to be authenticated. The UE assigns SID as the short-term context identity and derives session keys.

- $f4_{dhs}(SID, CID, AC, (SN \rightarrow UE)) \rightarrow sk_{SN}$
- $f4_{dhs}(SID, CID, AC, (UE \rightarrow SN)) \rightarrow sk_{UE}$

THEN: UE $\rightarrow$ SN: $\{M5(SID, \{RES_{HE}, CID\}_{cck_{UE}})\}_{cak_{UE}}$

6. *Message M6*

SN receives and decrypts **M5**. SN knows that SID is fresh. SN therefore has confirmation that the UE knows *dhs* and TI_{CA}. SN consequently accepts CID as an authenticated identity.

The UE and the SN may start using the short-term context $(STC_{(SN,UE)})$ and the session key set *sk* from this point in time. The SN has yet to receive the final confirmation from the HE, but the HE has verified that **M1** originated with the UE. The only thing the HE really needs at this stage is confirmation that the UE request (in **M1**) was fresh. The SN knows that the UE is online since the UE was able to respond with message **M5**.

THEN: SN $\rightarrow$ HE: $\{M6(CID, \{RES_{HE}, CID\}_{cck_{UE}})\}_{bkey}$

7. *Message M7*

The HE decrypts the outer block and sees **M6**. The HE then decrypts the inner block and verifies that the RES_{HE} is the correct response.

Subsequent to this the HE has fully authenticated the UE and accepts full liability for CID. The acceptance is communicated to both UE and SN.

$$- MACcck_{HE}(BIND, \text{``}Accept\text{''}) \rightarrow ICV_{M7}$$

THEN: HE $\rightarrow$ SN: $\{\text{M7}(CID, \text{``}Accept\text{''}, ICV_{M7}\}_{bkey}$

8. *Finally*

Subsequent to receiving **M7** the SN has full confirmation of HE liability for the UE (CID).

The SN then forwards the final confirmation. The final confirmation may be an information element added to a normal message on the CA channel.

THEN: SN $\rightarrow$ UE: $\{\text{Msg}(\text{``Message Contents''}, ICV_{M7})\}_{sk_{SN}}$

Prior to receiving the final confirmation the UE had not corroborated that HE believed CID.

$$- MACcck_{HE}(BIND, \text{``}Accept\text{''}) \rightarrow XICV_{M7}$$
$$- XICV_{M7} == ICV_{M7}$$

Table 7.6 Derived/learned UE knowledge.

Context	Data	Derived/Learned	Security Requirement
$MTC_{(HE,SN,UE)}$	VP	Computed	Non-Public;
$MTC_{(HE,SN,UE)}$	$SNID$	Broadcast from AP	Public;
$MTC_{(HE,SN,UE)}$	CID	Constructed	Subscriber Privacy;
$MTC_{(HE,SN,UE)}$	cck	Computed	Secret (between HE and UE);
$MTC_{(HE,SN,UE)}$	dhs	Decrypted, from **M4**	Secret (received from HE);
$MTC_{(HE,SN,UE)}$	cak	Computed	Secret (between SN and UE);
$STC_{(SN,UE)}$	CID	$MTC_{(HE,SN,UE)}$ binding	Subscriber Privacy;
$STC_{(SN,UE)}$	VP	$MTC_{(HE,SN,UE)}$ binding	Non-Public;
$STC_{(SN,UE)}$	SID	From SN, in **M4**	Sub. Privacy; Non-Public;
$STC_{(SN,UE)}$	sk	Computed	Secret;
$STC_{(SN,UE)}$	AC	Defined by AN/SN	Public;

Medium-Term Context and Short-Term Contexts (Post Conditions)

This is a summary of the derived/learned knowledge/state subsequent to executing the PE3WAKA_B2 protocol.

Table 7.7 Derived/learned SN knowledge.

Context	Data	Derived/Learned	Security Requirement
$MTC_{(HE,SN,UE)}$	VP	Received in **M1**	Non-Public;
$MTC_{(HE,SN,UE)}$	$SNID$	Self	Public;
$MTC_{(HE,SN,UE)}$	CID	Received in **M3**	Subscriber Privacy;
$MTC_{(HE,SN,UE)}$	dhs	DH exchange with HE	Secret (SN and HE/UE);
$MTC_{(HE,SN,UE)}$	cak	Computed	Secret (between SN and UE);
$STC_{(SN,UE)}$	CID	$MTC_{(HE,SN,UE)}$ binding	Subscriber Privacy;
$STC_{(SN,UE)}$	SID	Constructed	Sub. Privacy; Non-Public;
$STC_{(SN,UE)}$	sk	Computed	Secret;
$STC_{(SN,UE)}$	AC	Defined by AN/SN	Public;
$STC_{(SN,UE)}$	VP	Received in **M1**	Non-Public;

Table 7.8 Derived/learned HE knowledge.

Context	Data	Derived/Learned	Security Requirement
$MTC_{(HE,SN,UE)}$	VP	Received in **M2**	Non-Public;
$MTC_{(HE,SN,UE)}$	CID	Received in **M2**	Subscriber Privacy;
$MTC_{(HE,SN,UE)}$	cck	Computed	Secret (between HE and UE);
$MTC_{(HE,SN,UE)}$	dhs	DH exchange with SN	Secret (SN and HE/UE);

7.5.3 Security Argument

Security Goals

The security goals can informally be summarized as:

- *Entity Authentication*
  A main goal is to establish the medium-term context identity CID as an authenticated identity for all three entities. In line with the protocol design this is interpreted to be equivalent to a 3-way mutual entity authentication on CID.

    - *Mutual Entity Authentication between HE and UE*
      CID must be mutually authenticated between the HE and the UE.
    - *Mutual Entity Authentication between HE and SN*
      The HE and SN are already mutually authenticated, but this is not in itself sufficient. CID must also be mutually authenticated between the HE and the SN.
    - *Mutual Entity Authentication between SN and UE*
      CID must be mutually authenticated between the SN and the UE.

- *Confidentiality/Privacy*
  - CID shall only be known to the HE, SN and UE.
  - SID shall only be known to the SN and UE.[8]
  - The medium-term context shared secret, dhs, shall only to be known to the UE, SN and HE
  - Key set cck shall only be known to the HE and UE.
  - Key set cak shall only be known to the SN and UE.
  - Short-term context key set sk shall only be known to the SN and UE.

Brief Security Argument

The following is observed: The UE and HE is mutually authenticated with a dual challenge-response mechanism. The CID, as well as $UEID$, $SNID$ and $HEID$, is included as input in the response function. It is noted that CID is required to be a unique identifier with respect to the binding of the two challenge-response instances. Subsequent to the dual challenge-response the HE and UE will mutually believe in CID as an authenticated identifier.

It is a prerequisite that the HE and the SN are mutually authenticated prior to PE3WAKA execution. Furthermore, it is assumed that there exists a protected channel between the HE and the SN. The SN also assumes that the HE has jurisdiction over the UE. Thus, when the SN receives the CID over the protected CB channel (in message $M3$) it can tentatively assume CID to be a valid reference identity to the UE.

The mutual authentication between the UE and SN is somewhat complex. The following is noted: UE knows that CID is fresh. UE also knows that CID is only communicated in an encrypted block which is only intelligible to the HE. The HE has security jurisdiction over the UE and thus the UE trusts the HE in this respect. The UE therefore has assurance that the SN (with identity $SNID$, included in ICV_{M1}) has access to CID, which only could happen because it was authorized by the HE. When the UE receives block B from the HE it has assurance that the HE has acknowledged the SN and that furthermore the UE has received dhs, by which it knows that HE has designated dhs to be a shared secret with the SN. Then the UE uses the dhs to decrypt and verify that CID was available to the SN. This proves that SN has access to both dhs and CID, which is proof that the HE has authorized the SN. The UE then considers the SN to be authenticated and that furthermore

---

[8] This requirement only applies to the PE3WAKA protocol run. Subsequently, the SID will be used in plaintext.

that CID is recognized by the SN. The SN can correspondingly consider the UE to be authenticated on the CID when it receives message M5. The SN then knows that the entity presenting itself with CID was recognized by the HE and that the corresponding secret dhs is bounded to CID. The SN knows that the entity which replied with SID, encrypted with cak, must be the same entity which identified itself with CID. However, even if the SN already has authenticated the UE it has not yet received final confirmation from the HE. This arrives in message M7.

7.6 PE3WAKA Efficiency Analysis

7.6.1 Round Trips

Generic PE3WAKA vs. UMTS AKA

The following is a brief comparison of the round-trip efficiency of the PE3WAKA protocol and the corresponding 3GPP scheme. This amounts to assessing the round-trip cost of a UMTS location updating sequence including a UMTS AKA sequence and comparing it to the PE3WAKA round-trip cost (see 3G TS 23.108 [165, section 7.3.1])):

- UE→SN: *Location Updating Request*
- SN→UE: *Authentication Request*
- UE→SN: *Authentication Response*
- SN→UE: *Location Updating Accept*

The 3GPP location updating sequence is essentially identical to the 2G (GSM) sequence. The TS 23.108 specification only covers the UE-SN communication. The SN-HE part can be found in TS 29.002 [58], and it constitutes two separate request-reply sequences where the SN first fetches the subscriber information and then the security credentials.[9]

In the PE3WAKA protocols the subscriber information will be forwarded in parallel with message M3. The PE3WAKA protocols perform slightly better than the UMTS scheme on the UE-SN interface. Note that message M4 indirectly serves as location registration confirmation. On the SN-HE interface, the

---

[9] The main MAP procedures of interest are the UPDATE_LOCATION and the INSERT_SUBSCRIBER_DATA procedures. These must always be executed. Additionally, one may have to execute SEND_INDENTIFICATION (request IMSI and AV from previous VLR) and CANCEL_LOCATION (to cancel registration with old VLR). If the SEND_INDENTIFICATION procedure was not executed then the VLR probably has to execute the SEND_AUTHENTICATION_INFO procedure. There are even more additional procedures, but these are either local (between VLR and MSC) or entirely optional.

PE3WAKA protocol requires two passes. The UMTS scheme also requires multiple messages to be exchanges, but the PE3WAKA protocols may have a slight advantage here.

It is observed that the UE-SN context is potentially operative after SN reception of message #5. The SN still wants HE confirmation, but it now has indirect CID confirmation. It is therefore safe for the SN to activate the short-term context (with a pending confirmation). One can therefore defend the claim that PE3WAKA is at least as efficient as the comparable UMTS sequences.

7.6.2 PE3WAKA Message Sizes

A concern for the PE3WAKA protocol has been the capacity of the (radio) common channels of the A-interface. There are two main cases to be considered for the PE3WAKA protocols: with or without DH over the A-interface. Executing the DH exchange over the CA channel is potentially problematic.

The B-interface is a fixed line interface. In this context the capacity requirements of the PE3WAKA protocol is modest and there is no reason to expect the B-interface to be a capacity bottleneck.

It is noted that counts below only include the bare bones information elements. The encoding overhead is not included. In practice that may account for 1-3 bytes per IE for flexible coding schemes. Additionally, IEs to denote protocol version, message identifiers, etc., are not included. The overhead here is relatively small, but should still be taken into account for severely bandwidth limited environments. It is also noted that algorithm negotiations may be required and that algorithm indication/algorithm requirement is necessary for proper binding. None of this is include in the example protocol.

Message Sizes

We only investigate the CA channel here since it is the potential bottleneck in the protocol execution. The information element bit-sizes given below is only an estimate:

- **M1:** The M1 message contains one public-key encrypted block and several other IEs.
  - The size of Block A is expected to be identical to public-key block size. Block A contains $UEID$ (max. 128 bits), CID (64–128 bits), CH_{UE} (128 bits) and ICV_{M1} (64 bits), which easily fit inside one

block. The block size is nominally set to 1500 bits, but may be larger (or smaller for ECC based systems).
- The identity $HEID$ is expected to be 128 bits wide (or less).
- The validity period VP is expected to 64 bits wide (or less).
- Some additional protocol housekeeping overhead must also be added.

$|M1| \leq 2048$ bits

- **M4:** The M4 message consists of data from both HE and SN.

    - Block B contains a challenge (128 bits), a response (64 bits) and CID and the DH secret (256 bits).
    - The block with SID allocation consists of 1 to 2 128-bit wide blocks.
    - Some additional protocol housekeeping overhead must also be added.

$|M4| \leq 1024$ bits

- **M5:** This message contains response data to the SN/HE.
  The block size of the symmetric cipher is expected to be 128 bits wide. To cover this 2–3 blocks may be necessary.
  $|M5| \leq 512$ bits

- **Mx:** This message contains the final confirmation.
  One 128-bit block should suffice here.

We do not foresee any particular problem with the above quoted message sizes. If one chooses to execute the DH exchange over the CA channel the issue gets more complicated. It may be problematic to initiate a sequence with DH parameters in the order of 15–16 Kbits over the CA-channel. Using an elliptic curve version of the DH protocol (ECC-DH) may help considerably and may make the approach feasible, but for performance/capacity reasons it seems best to execute the DH exchange over the CB-channel.

7.6.3 Computational Aspects

It is not expected that the use of symmetric methods (encryption/decryption, key derivation, MAC computation) will impose a significant workload. There will be some overhead, but it will be relatively small compared to the public-key operations. Thus, only the public-key operations need really be taken into account.

Instantaneous Demand

It is assumed that the DH group parameters is agreed *a priori* between HE and SN (one may of course occasionally change the parameters). The SN and HE may therefore precompute a set of DH public keys and store them for usage when a registration event happens. The instantaneous demand can therefore be reduced to:

- **UE**: One public-key encrypt operations ($M1$).
  The UE initiates $M1$, and the real-time requirements here are tied to radio environment conditions and user perceived performance.
- **SN**: The SN must compute one DH parameter ($M2$) and DH secret.
  The DH parameter must be available before $M2$ can be sent. The SN must compute the DH shared secret after receiving message $M3$ and before transmitting message $M4$.
- **HE**: The HE must execute DH operations and one public-key decrypt.
  The HE must decrypt block A ($M2$) before it can progress with $M3$. HE must also compute a DH parameter and the DH shared secret before transmitting $M3$.

The fixed network nodes (HE/SN) are powerful entities and can easily carry out the required computations. Still, the HE and SN nodes must support a high number of UEs. It may therefore be beneficial to have dedicated hardware to execute the public-key and DH computations. Overall, the HE and/or SN should not have any problems with the computational burden imposed by this PE3WAKA protocol.

The UD/SM is also a relatively powerful computational platform these days. The UD/SM, moreover, can dedicate all its resources to the setup. Today, smart cards are able to execute asymmetric primitives with relative ease [163]. For our example protocol the DH exchange is not executed on the UD/SM, but for some PE3WAKA protocols this may be the case. The requirements on the UE (one public-key encrypt and potentially DH computations) are tough compared to the 3GPP AKA. The SM platform must therefore be relatively powerful. Nevertheless, it should be feasible having the UE carry out one public-key encryption and (potentially) one DH exchange per PE3WAKA invocation.

Total Demand

The total computational demand of the PE3WAKA protocols is significantly higher than for the symmetric-only 2G/3G AKA protocols. However, the demand of the 2G AKA, when it appeared around 1990, was considerably tougher to meet by the smart cards available at that time. We therefore see no compelling reason why the PE3WAKA protocol cannot be supported by a next generation wireless/cellular system.

The SN nodes may have to serve several hundred thousands of subscribers. In all likelihood one would need dedicated crypto accelerator hardware; such hardware is commercially available and so the SN nodes should be able to meet the needs. Individual PE3WAKA invocations can easily be processed in parallel, and thus scalability is not a big issue.

The HE nodes may serve even more subscribers than the SN. This may require a larger pool of crypto accelerator hardware, but again the problem should be easy to contain.

7.7 Other Aspects

7.7.1 Denial-of-Service (DoS)

The PE3WAKA protocol does not provide explicit DoS protection. The PE3WAKA protocol operates in an environment where it is very easy for an adversary to carry out access-denial DoS attacks simply by disrupting the radio transmission. Access-denial attacks are local in nature and since they do not scale we have not tried specifically to avert this type of attack in the PE3WAKA protocol. To limit computational DoS attacks we suggest that the SN restricts the arrival rate of PE3WAKA invocation per access point. The HE, likewise, may limit the number of simultaneous PE3WAKA sessions from any given SN network element. Together, this should effectively prevent a computational DoS attack from scaling.

7.7.2 Initiator-Responder Resilience

The PE3WAKA protocol does not specifically aim at providing initiator or responder resilience [111], but informal analysis suggests that it is difficult for any principal to gain any advantage here.

DH Exchange between SN and HE
The UE does not influence the dhs, but it selects the CID. The CID is used for key derivation and thus the UE has an influence on (all) the session keys. The SN does not know the CID when it generates the DH_{SN} parameter, so its influence is contained on the medium-term context. However, the SN will choose the SID and will thus influence the session keys. The HE will know both CID and DH_{SN}, and it may potentially tailor a DH_{HE} to create a

specific dhs. However, to control the session keys (stk) one must also know SID, and HE will not normally know SID. So the HE cannot in practice control the session keys.

DH Exchange between SN and UE

The UE will influence the dhs and it selects the CID. The CID is used for key derivation and thus the UE has an influence on (all) the session keys. The SN does not know the CID when it generates the DH_{SN} parameter, so its influence is contained on the medium-term context. However, the SN will choose the SID and will thus influence the session keys. The HE will have rather limited influence on the key sets unless special care is taken in the key derivation functions.

For the example protocol the practical HE influence is almost none on the session keys, but the UE and SN are evenly matched.

7.7.3 Is the DH Exchange Really Needed?

For the purpose of PE3WAKA the choice has been to base all protocols on a DH exchange. However, one can easily construct a shared secret that is *not* based on a DH secret. For instance, one could have constructed the medium-term context secret from a MAC/hash-based derivation over the CID and the challenges. Computationally this would have been simpler and one would not have needed to exchange the DH public parameters.

It is nevertheless argued that the use of a DH procedure is preferable in that one has a very solid basis for the derivation of the shared secret. Still, it is acknowledged that one may conceivably avoid the DH exchange entirely if one is prepared to have a slightly less solid foundation for the shared secret. Also, the use of a DH exchange is better with respect to perfect forward secrecy (PFS), which is commonly considered a desirable property a protocol to have. In practice we believe that the advantage of PFS should not be overestimated for an AKA protocol aiming at securing access security.

7.8 Guideline Compliance

The guidelines outlined in Chapter 6 need not be followed to the letter, but one should have good reasons for not adhering to the principles.

7.8.1 Concerning the Prudent Engineering Principles

Several steps have been taken to conform with the prudent engineering principles. Amongst those are explicit use of nonces and explicit use of cryptographic mechanisms. For instance, the CID could additionally have served as the UE challenge data, but the design would then have hid that there are separate duties to be performed. Furthermore, the use of both an explicit challenge-response mechanism and the extensive use of integrity protection was done to separate the task being achieved.

The only prudent engineering principle that may be said to have been violated is the principle which states that "It should be possible to deduce the protocol, which run of the protocol and the message number within the sequence by inspecting a message." (**PE.9**).

As has previously mentioned, the proposed example protocols would need additional work to be transformed into real-world operational proto-cols. In particular, the encoding of the IEs must be formalized. This would naturally also extend to devising encoding schemes for message numbers and message formats as well as detailing the protocol identifiers and the protocol version numbering.

7.8.2 Concerning the Cellular Access Security and Privacy Guidelines

The specific guidelines proposed in this thesis

1. *Home Control*
   The home control requirement is mainly satisfied through the online 3-way architecture of the PE3WAKA protocols.
2. *Online Security Context Establishment*
   All PE3WAKA protocols are online for all three parties with respect to the medium-term security context.
3. *Active 3-Way Security Context Establishment*
   All PE3WAKA protocols requires active participation by the principals.
4. *3-Way Registration Control*
   This is true for the PE3WAKA protocols.
5. *Security Context Hierarchy*
   The PE3WAKA protocols operate within the context of a 3-level secur-ity context hierarchy. The number of separate context may vary, but the basic 3-level hierarchy structure is present for all PE3WAKA protocol variants. As is the case in LTE, chapter 3, one may also need to sep-

arate the short-term contexts into *control plane* and *user plane* specific context.

6. *Security Context Exposure Control*

   The exposure control is enforced through the binding to both the area (the area code) and through a validity period. Exposure is thus limited to the confines of the defined area and the defined period of time. Additionally one may limit the number of derived sub-context, etc., but this has not been investigated or actively considered.

7. *Security Context Separation*

   This issue is not addressed in the proposed example. Indeed, the issue is outside the scope of the example. However, the PE3WAKA protocols will easily cater for different contexts for different access types, but this needs to be explicitly encoded. One must also decide whether or not separate context hierarchies are needed or if the short-term contexts may be based on a common medium-term context. The medium-term context *may* be shared, but the safest practice seems to be only to share the long-term context.[10]

8. *Spatio-Temporal Security Context Binding*

   This requirement is in practice quite similar to the exposure control requirement. The PE3WAKA protocol supports this principle.

9. *Security Algorithm Separation*

   The PE3WAKA scheme does not directly/explicitly support this. The security algorithm negotiations and subsequent binding must be integrated in the protocols before they can be considered operationally complete. As was highlighted in Section 4.7.4 the separation should also cover mode-of-operation separation. The meta-data required to support this could usefully be included in the expiry conditions of the key. The algorithm binding may be static, at the creation of the key, or dynamic. Dynamic binding would then be done to the algorithm-choice signalled in the `security-mode-command`.[11]

10. *Key Sets and Key Strength*

    The key strength requirements are supported by the PE3WAKA protocols, but the actual requirements may vary and so the key derivation algorithms, must be configured to accommodate the specific needs.

11. *SN Security Termination*

---

[10] This is what is being done in 3GPP, where the USIM/HLR long-term credentials can be used as a basis for EAP-AKA derivation of 3GPP-WLAN interworking session keys, for derivation of IMS session keys, etc., in addition to being the basis for the UMTS session keys.

[11] A command to signal start of ciphering/integrity protection on a bearer channel.

This issue is somewhat outside the scope of the generic PE3WAKA design. However, the issue is real and it needs to be carefully addressed for any operational system. The decisions may result in a mismatch between key derivation point and the key usage point. Any ensuing key distribution problems must be resolved. One may also need to have a two-layered structure as is used in LTE (see Chapter 3), and then there may be several termination points.

12. *Identity Management, Registration and Security Setup*
    The PE3WAKA protocols are designed to provide this integration.

13. *Channel Allocation and Session Key Agreement*
    This issue is not directly addressed by the PE3WAKA proposals as the PE3WAKA protocols are not intended to be used for reestablishment of the short-term security contexts. The reestablishment protocol may be implemented as a stand-alone protocol or it may be included as a component in relocation procedures or similar.

14. *Subscriber Privacy – Domain Separation*
    The PE3WAKA protocols adhere to the principle of domain separation between HE and SN domains, but it is observed that no AKA protocol can enforce this principle. The enforcement must be through mandatory separation of HE and SN in term of operative and administrative management.

15. *Subscriber Privacy – Identity Privacy*
    The identity privacy scheme employed by the PE3WAKA protocols provides credible identity privacy both against the external intruder and against internal entities. Only the HE and UE will ever know the $UEID$ and the CID is only visible to the UE, SN and HE. The SID is nominally only know to UE and SN, but will be used for paging and access request purposes.

16. *Subscriber Privacy – Location Privacy*
    The PE3WAKA protocol provides a measure of location privacy, and in particular it ensures that the binding between any observed position (x, y) and the identity of the entity at position (x, y) is not revealed.

17. *Subscriber Privacy – Untraceability*
    Ultimately, this is an impossible goal for an entity which communicates over a radio channel. However, steps are taken to limited the potential for tracking and the usage of the SID is important here. It is noted that the SID replacement frequency will affect the degree of "Untraceability".

18. *Subscriber Privacy Override*

The Lawful Interception and the Emergency requirements (E112/E911) for privacy override can be adhered to if the HE and SN cooperates.

There are some requirements that are not currently catered for by the example PE3WAKA protocols. This can be explained by the fact that these are example protocols and not fully defined operational protocols. There are also some requirements that fall slightly outside the scope of the AKA protocols. However, to the extent possible and practical the guidelines have been adhered to.

7.8.3 Concerning the System Evolution and Management Guidelines

The management guidelines are hard to assess fully in the context of example protocols. Indeed, these principles are intended for the design of the full system and security architecture.

1. *Design for System Evolution*
   No explicit algorithm negotiation or algorithm deprecation mechanisms have been designed for the PE3WAKA example protocols. Clearly, algorithm indication, algorithm negotiation, algorithm binding and algorithm deprecation must be implemented for a fully defined operational protocol to fulfill this requirement.
2. *Backwards Compatibility*
   The PE3WAKA protocols are indeed not designed to feature backwards compatibility with any earlier protocols.
3. *Compromise Impact Control*
   A measure of impact control is designed into the PE3WAKA protocols. Compromise of one UE should not affect any other UEs unduly. Complete compromise of HE and/or SN obviously constitutes a major breach of security, and this cannot be mitigated or prevented by the PE3WAKA protocols.
4. *Never Willfully Design "Weak" Versions of Algorithms or Protocols*
   The PE3WAKA protocols do not provide support for any "weak" parts. Still, the very fact that (by explicit requirement) subscriber privacy is revocable by authorized request from the regulatory actors means that there is a weakness in the privacy mechanism. However, this is the result of an explicit design requirement and must thus be permitted.

7.8.4 Guidelines Summary

The PE3WAKA example protocol does not fully comply with the principles and design guidelines. Some of the guidelines are outside the scope of what a PE3WAKA protocol can directly address. These issues should be addressed in the context of the overall system architecture and the context of the complete security architecture. Some of the principles and guidelines target implementation issues not addressed in the example protocols. These must be addressed for a fully operational protocol.

The principles and guidelines have proved to be somewhat overlapping in practice, but it has been clarifying to design example protocols for each guideline independently. It has also been a good exercise to evaluate the example protocols for compliance in that it has highlighted what is missing if the example protocols are to be converted to fully defined protocols.

7.9 Summary

In this chapter it has been demonstrated that one can design effective and efficient privacy enhanced 3-way authentication and key agreement protocols for future mobile systems. The example PE3WAKA protocol presented is capable of substantially improved user location/identity privacy compared to the 3G scheme. The PE3WAKA protocols provide enhanced privacy from eavesdropping and manipulation by an outsider adversary and provide a measure of user location/identity privacy with respect to the serving network and the home operator. To achieve this the user identity presentation scheme and the initial registration procedures had to be modified from the scheme used in 2G/3G systems. However, this change also allowed performance improvements by integration of signalling procedures that are triggered anyway by the same physical events. The PE3WAKA protocols also recognize that all three principals should participate in the security context, and they provide online establishment of a 3-way security context. Finally, the PE3WAKA protocol provides enhanced key distribution relative to the 2G/3G schemes. In the PE3WAKA protocol the serving network participates actively in creating security credentials for session key derivation. Contrast this with the 2G/3G scheme in which the home entity unilaterally decides the session keys.

The design of the PE3WAKA example protocol was done in the context of the design guidelines proposed in Chapter 6.

It is noted that although a brief security arguments have been given for the example protocols, they should nevertheless be subjected to a formal analysis/verification.[12] Formal verification of security protocols is essential, although not so much for proving the protocol correct as for ensuring that the most awkward design errors are avoided and for clarifying the design. Formal analysis is the topic of the next chapter.

---

[12] Several PE3WAKA example protocols have been successfully verified with the AVISPA tool set [166].

8

Security Protocol Verification

> Program testing can be used to show the presence of bugs, but never to show their absence!
>
> *– Edsger W. Dijkstra (1930–2002)*

8.1 Security Verification - Methods and Tools

8.1.1 Perspectives on the use of Formal Methods in Proving Security Protocols

Gollmann [167] brings some perspective to the process of formal verification and on the five "promises" of formal analysis (from [167]):

- Formal modelling makes security properties precise.
- Formal modelling makes security properties clear.
- Formal specification makes security protocols precise.
- Formal methods can deal with the complexities of security analysis.
- With formal methods we can prove security.

The "promises" are then discussed, and while it is apparent that the use of formal methods has advanced the field and made it possible to construct better and more secure protocols, it is also apparent that the "promises" are not fully kept. Amongst other things Gollmann notes that precision and clearness may be somewhat contradictory goals. There are also serious issues with protocol execution context and security assumptions (like trust, and so on). When it comes to dealing with complexity there has been much progress. This has allowed the formal methods to move on from "toy" protocols to complex real-world protocols (like IKEv2 [138] or for that matter the PE3WAKA protocols). It also allows for models that are in better correspondence with the target protocols.

Proofs are described by Gollmann as a "non-goal". Proving specific properties under specific circumstances is obviously possible. Nevertheless, this does not "prove" the protocol.

8.1.2 Some Efforts in Security Protocols Verification

Some of the PE3WAKA protocols have successfully been verified with the AVISPA tool set [166]. The AVISPA tool set has the advantage of being a recent effort and being a quite well developed set of tools and methods. Furthermore, the AVISPA tools are being actively maintained and it has an active user community. The AVISPA tool set will be presented in more detail in the next section.

The AVISPA framework is not the only game in town. Below is a brief presentation of some works in the area of formal methods for security protocols, including a few proposed languages, tools and frameworks. The presentation given below is by no means complete, but hopefully it is a relatively representative presentation of the research that has been ongoing in this field the last 20 years. The Boyd–Mathuria book [121, chapter 2]) provides a somewhat more complete exposition of the various formal verification efforts (up to around 2002).

8.1.3 BAN Logic

BAN logic first appeared in the 1989 DEC SRC report no. 39 "A Logic of Authentication" [66] by Burrows, Abadi and Needham. The basic logic is a belief logic. A principal *sees* what it receives. Conditional to cryptographically based assurance (explicit logic rules), the principal may believe that another principal *once said* the message element. If it furthermore can be proven that the message element is *fresh* then the principal is entitled to *believe* in the message element. Additionally, cryptographic key assurance may require that a principal has *jurisdiction* over the key.

The logic, almost universally referred to as BAN logic, is no longer the preferred method for deriving security proofs. The field has matured, progressed and moved on beyond BAN logic. Automated tools for BAN-like logics were developed but have now been discontinued; an example is given in [65].

8.1.4 Spi-Calculus

In [168, 169] Abadi and Gordon introduced the spi calculus. The spi calculus is an extension of the pi calculus; The pi calculus being a powerful process oriented calculus with an explicit channel notion. The spi calculus was designed for describing and analyzing cryptographic protocols. The spi calculus therefore has extensions for representing encryption/decryption operations and considers cryptographic properties directly. A specific goal was to use spi calculus for studying authentication protocols.

It is noted that the capabilities and properties of the intruder need not be coded or specified in the spi calculus models. This is both a strength and a weakness; the strength being that one need not encode specific attacks or intruder models, which often is a tedious and error prone process. Correspondingly, the weakness is that one cannot directly qualify the strength of the intruder. The issues here are similar to other formalisms and methods which assume a DY type of intruder. It is also noted by the authors that "Proofs in the spi calculus are sometimes more difficult than proofs in those earlier frameworks" (in [170, section 8]). While not a fault of spi calculus *per se*, it does illustrate that the spi calculus is hard to use manually. Thus, it was recognized that for widespread use the proof process would have to be automated.

8.1.5 FDR/Casper

The FDR/Casper tools are based on the Communicating Sequential Processes (CSP) process algebra developed by Hoare [171, 172]. CSP itself does not deal with security properties, but is rather a generic language in which to model asynchronous processes and protocols. The Failures-Divergences Refinement (FDR) tool is a model checker for CSP. A companion to FDR is the Casper tool. Casper is basically a compiler of security protocol models and the derived models can subsequently be verified/falsified with FDR. FDR and Casper are documented in depth in the book *The Modelling and Analysis of Security Protocols: The CSP Approach* [173] by Ryan and Schneider. In the paper [149] Lowe investigates the use of FDR in analyzing security protocols. In a related paper [174] the author takes advantage of using Casper to specify the security protocol model. This has the added advantage that Casper automates the description of the generic intruder.

8.1.6 The NRL Analyzer

The Naval Research Laboratories (NRL) Analyzer [175] is a tool/language developed by Meadows. The basic approach is that of a theorem prover and the NRL Analyzer is written in Prolog. The specification contains transition rules and stated facts. Interestingly, the NRL Analyzer has a construct to explicitly model intruder knowledge (the `adversaryknows()` statement). The Analyzer is based on a term-rewriting model of a DY intruder and equational unification. The choice of Prolog as the implementation language was seen as an clear advantage by the author, which states that Prolog afforded considerable flexibility in the design process. It is also noted that unification is a primary strength of Prolog.

8.1.7 Isabelle/HOL

Isabelle is a proof assistant tool verifying/falsifying proofs developed in Higher Order Logic (HOL). The Isabelle tool is implemented in ML. One of the benefits of using HOLs is that one directly will have type verification, and as was noted in [139] type errors are a common source of protocol errors. Isabelle/HOL is not directly aimed at verifying the correctness of security protocols *per se*, but it has been demonstrated that Isabelle/HOL can be successfully applied to this domain. In [176] Paulson develops a set of security theorems (very loosely built on BAN logic) in HOL and then proceeds to verify/falsify example protocols (variants of the Otway–Rees mutual authentication protocol [177]).

It is observed from Paulson [176] that deriving security protocol models in Isabelle is far from an easy exercise and that the correspondence between the model and the target protocol tends to be non-obvious.[1] That in turn may cast serious doubts on the *validity* of the obtained results. This problem does not only affect Isabelle/HOL, but is a generic problem for formalisms in which the correspondence between the model and the target protocol is non-obvious.[2]

---

[1] That is (in line with Gollmann [167]); one may very well have *precision*, but *clarity* and correspondence may be lacking.

[2] The problem is also present for the cases when the abstraction level is too high; i.e. when too many of the aspects of the target protocol are not expressed in the model.

8.1.8 Brutus

The paper [178] describes the Brutus security model checker tool. The Brutus modeling language is logic based, using a specialized first-order logic with quantifiers and past-time model operators (i.e. to describe events and protocol runs in the past). The paper emphasizes that verification tools should be easy to use[3] in order to have practical application. This goal is clearly important to gain widespread use outside the formal methods/security research community.

8.1.9 Action Language for Security Protocols

Sometimes the most productive approach is to turn the issue upside-down. This is the basis for the work on $\mathcal{AL}_{SP}$ (Action Language for Security Protocols) approach. The logic is used for "planning attacks to security protocols" [179]. A failure to find an attack is equivalent to proving that the protocol model is safe within the constraints of the logic. The "attack plans" in question are protocols traces and the actual execution of the logic is based on model checking methods. The comprehensive paper "Verifying Security Protocols as Planning in Logic Programming" [179] is a good reference the $\mathcal{AL}_{SP}$ approach.

8.1.10 Symbolic Trace Analyzer (STA)

The Symbolic Trace Analyzer (STA) [180] is a model checking tool with an emphasis on the *symbolic* part. That is, by using lazy evaluation techniques it avoids construction of the full state-space and this allows the STA tool to be both effective and efficient. One can then avoid some of the state-space explosion problems normally associated with model checking methods. The STA tool is implemented in ML and the language used for modeling is a variant of the spi-calculus.

8.1.11 Spin/Promela

The Promela language and the Spin/XSpin model checking tools are not directly targeted at proving (or rather: prove absence of contradictions) security properties. As such there are no primitives in Promela for expressing security properties and conditions.

---

[3] "Therefore, in order for a tool to effortlessly blend into the software process a verification tool needs to be lightweight and push button in nature" [178].

Nevertheless, Spin/Promela has been used for verifying security protocols and since it is a highly optimized framework performance-wise it is also an efficient verification tool. The primary reference for Spin/Promela is the book by Holzmann [67].[4] In his PhD thesis [181] Ruys identifies several strategies for achieving efficient verification in Spin/Promela. There have been dedicated efforts to model security properties in Spin/Promela [182–187]. Some of these papers are quite interesting, but verification of security properties is nevertheless not a primary strength of the Spin/Promela model checking tools.

8.1.12 The AVISPA Tool Set

The AVISPA[5] tools and methods are developed within the context of the AVISPA IST project. The AVISPA tool set is aimed at "... developing a push-button, industrial-strength technology for the analysis of large-scale Internet security-sensitive protocols and applications."[6] A good introduction to AVISPA is found in the AVISPA Tutorial [188].

The High Level Protocol Specification Language
Originally within the AVISPA project one attempted to design the modelling language (HLPSL) around an augmented Alice-Bob notation, but found that this imposed too many restrictions. Particularly, it was difficult to encapsulate concurrency and conditional execution in the Alice-Bob notation. Semantically, HLPSL is based on Lamport's Temporal Logic of Actions (TLA) [64], but the HLPSL language has a number of syntactical constructs and abstractions that allow the modeler to express the security properties more directly.

Composition of HLPSL Models
Protocol specification is HLPSL is organized in a set of basic roles and a number of active agents that can instantiate the roles. The basic roles are composed in session roles (protocol run sessions). The session roles are again initiated in an outer role (the environment), where one defines the environment context of the protocol run. The HLPSL language directly supports the notion of an intruder, specifically the Dolev–Yao intruder [100]. By definition

---

[4] Spin/Promela is actively maintained. See *www.spinroot.com* for up-to-date information on Spin/Promela.

[5] Automated Validation of Internet Security Protocols and Applications (AVISPA).

[6] Quote from the AVISPA homepage (http://www.avispa-project.org/).

the HLPSL language assumes that the network is the intruder. Additionally, the intruder can be set up to play basic roles (simulating a dishonest principal) and to be given explicit knowledge (which may be used to simulate the resilience of the protocol in the face of exposed credentials).

Verification

The HLPSL models are translated and compiled into an intermediate format. The compiled model is then verified by one of the AVISPA back-end verifiers.

- On-the-Fly Model-Checker (OFMC)
  The OFMC back-end is implemented in the Haskell language and is principally due to Mödersheim. The OFMC tool is documented in considerable detail in Mödersheim's PhD thesis [189].
- Constraint Logic Attack Searcher (CL-AtSe)
  The CL-AtSe back-end is based on a lazy intruder technique similar to that of OFMC, but unlike OFMC it does not incorporate constraint differentiation. CL-AtSe is documented in [190].
- SAT Model-Checker (SATMC)
  The SATMC (SAT Model-Checker) tool is based on encoding the model (the problem) into a Boolean formula. If the formula is satisfiable then there is an attack. The SATMC tool is documented in [191, 192].
- Tree Automata for Security Protocols (TA4SP)
  The TA4SP tool is based on the use of tree-automata to represent an over-approximation of the intruder knowledge that can be reached. A benefit with this tool is that for error-free protocols it can be shown that the result is valid for an unbounded number of sessions. It has however some severe limitations, in particular that it cannot check authentication goals (only secrecy). This makes TA4SP considerably less useful. Use of TA4SP is documented in [193].

The back-end verifiers are not created equal. According to [188], only the OFMC and CL-AtSe supports exponentiation (the `exp()` operator). This is needed if one wants a faithful model for a DH exchange. Given this, only the OFMC and CL-AtSe back-ends are considered suitable for verifying PE3WAKA type of protocols.

The Security Protocol ANimator (SPAN) Tool

This tool is not directly part of the AVISPA tool set, but is designed to work directly with the AVISPA tools. The SPAN tool permits interactive design of

HLPSL models and is a most valuable tool in the protocol design process. SPAN is therefore an essential tool in the design and development of HLPSL verification models, making the power of the AVISPA system much easier to harness for people outside the AVISPA development community. The SPAN tool is documented in the user manual [194] and in the papers [195, 196].

8.2 Other Aspects of Formal Verification

8.2.1 What Exactly Constitutes a Security Breach

An important question to ask when analyzing the various formal approaches is to determine what really constitutes a successful attack. Obviously, a successful attack would have to meet some goal of the intruders, but quite frequently the goals of the intruder are different from breaking the expressed security goals. Also, the security goals are often vaguely and imprecisely stated. Adding to this, many of the attacks found can also be considered artifacts of the model and the formalism applied.

One must therefore investigate whether a given attack really achieves anything of substance. This, in turn, can only be evaluated in the execution context of the discussed protocol. As is frequently the case, the protocols are proposed in a context which makes some attack types utterly impractical or impossible. For instance, rarely is it the case that one truly confronts a DY intruder and for security protocols executed *only* as part of a larger protocol stack it may be quite irrelevant if a replay attack is possible for the stand-alone protocol (assuming that attack is not viable in the larger context).

There is also the issue of trust and honesty. Protocols often have an implicit assumption about whether there can be honest, semi-honest and/or dishonest principals. This does have a strong bearing on what exactly a security protocol can provide and when the results can no longer be trusted. These issues are discussed in [197] and the conclusions are interesting in that it questions the validity of several of the flaws found in security protocols.

8.2.2 Formal Methods as a Design Tool

Few of the formal verification tools are in close correspondence with the actual design or construction of protocols. The expression languages of the verification tools and the formal model of the target protocols tend not to be very helpful in designing a protocol. For a protocol designer it is of course of interest to know whether the protocol is safe or not, but this is not in itself of very much help in the actual design process. In the paper "Protocols Are

Programs Too: The Meta-Heuristic Search for Security Protocols" [198] this is explicitly recognized.[7] The authors propose and discuss a framework for automated protocol synthesis to help aid the protocol design process.

8.2.3 Counterexamples and Multiple Instances

The tools and methods discussed above have different properties when it comes to being a useful design tool. In particular, when a model has been falsified it is of great interest to se exactly why the model was falsified. The model checkers are generally good at providing counterexamples, i.e. pointing directly to an example of the falsification. With theorem provers, on the other hand, it is normally much more difficult to provide counterexamples. On the other hand, the problem that model checking tools tend to have with multiple protocol instances (due to state-space explosion) is not an issue for theorem provers. Adding to this, it is noted that many problems cannot be found without considering multiple protocol instances. So there are pros and cons to the theoretical foundation of the different formal methods used [178, section 7].

In this book the theoretical foundation of the tools is not of primary interest, but the usability of the protocol verifiers in protocol design is definitively an issue. So too is the scope of the methods with respect to what they can and cannot do.

8.3 Summary

8.3.1 Validity

Formal verification of security protocols has moved on from being only an esoteric activity carried out by academics to being a practical tool during development of high-confidence security protocols. The "development tool" aspects should be stressed.

One should also note that none of the formal methods work on the security protocols themselves. Instead, one defines a model of the security protocol. The model is formulated in some formal language and it, at best, represents a simplified version of the security protocol. However, quite frequently there are aspects which one cannot directly express in the formalism. For example,

---

[7] "Designers often wish to find an efficient way of implementing a specification and they may also want to explore the consequences of making different initial assumptions or even requiring different goals" [198].

in PE3WAKA a convoluted approach is used. A message may be composed of several parts and in order to decrypt part B one first needs to evaluate part A and derive the key to decrypt B. This implies a strict order in the processing of the message. In formalism like HLPSL this is not easily captured as all expressions within a state are instantaneously evaluated. Thus, part B is not allowed to wait for its key to be derived and consequently the AVISPA back-end analyzers will tell you that the message cannot be evaluated.

This highlights an important aspect of formal verification, namely that of validity. Does the expressed model truly represent the actual security protocol? It may not be a complete representation, but is it even a proper subset of the actual protocol? And, does this proper subset really express the properties that we need to evaluate? How can we know whether or not the model is a valid model of the actual protocol? The validity problem is compounded for security protocols in that security is a weakest link game and that therefore every simplification or modeling liberty is suspect by nature. How can we know if our simplification or modeling liberty did not reduce the whole model such that an actual attack is now prevented (in the model)? The answer is that we cannot know.

The only thing we can do is try to ensure that the conceptual distance between model and the target protocol is as small as possible. The AVISPA project started out with an Alice-Bob notation for the protocol modeling. Conceptually, this would be close to the way many protocols are defined, but there are concepts which cannot easily be captured this way (conditional expressions, etc.). So the AVISPA project devised HLPSL based on TLA. This permitted much more complex cases to be modeled. TLA is based on firm formal ground and so it seems everything is fine. That is, except that protocols are not designed like a HLPSL model is designed. In particular, the temporal model of TLA/HLPSL does not match well with how real security protocols are constructed and composed. One can certainly find ways around the problems, but this is beside the point since these work-arounds only increase the conceptual distance between the model and the target protocol. Thus, what was gained in terms of verification capabilities had a considerably cost in terms of lost assurance of validity.

8.3.2 Verification

Validity is not a binary concept in which a statement is either true or false. There will be degrees of validity, and so even if there are aspects which the model does not capture it may nevertheless be a useful model.

Having derived a model one must then clarify and express the assumptions. Within that context one can attempt to verify or falsify the model. In this area the tools and tool sets have matured significantly and many properties can be automatically checked. Entire subclasses of design faults can be captured this way. This is very useful and it will prevent the most obvious (with hindsight) and embarrassing errors.

However, there are limits to what the tools can do and a "successful" outcome does in no way prove the absences of errors. So, philosophically, one must use the verification process to catch as many errors as possible and not lull oneself into a belief that the use of formal verification tools can prove the protocol error-free.

8.3.3 The Simulation Aspect

The SPAN tool is a most valuable addition to the basic AVISPA tool set. The design advantages of being able to verify executability of the HLPSL models and to refine and debug the HLPSL models are significant. The model simulation environment of the SPAN tool therefore contributes significantly to the design process, and it is the experience of the author that it also contributes to the confidence one has that a model is a valid representation of the given protocol. Given that there is a certain conceptual distance between the actual protocol and the HLPSL model it is very useful to gain "visual" confidence that there is indeed correspondence between the designed protocol and the associated HLPSL model. This is in line with the observations made by the SPAN designers [195].

The simulation also provides the protocol designer with freedom to experiment during the design phase. This is most useful and it will likely improve the quality of the protocol in many ways, including improving the security.

9

Summary

Quis custodiet ipsos custodes? (Who watches the watchers?)

– Proverb

9.1 Challenges

There are many open issues in the area of access security and subscriber privacy for cellular systems. The following are just some of the open issues:

- Optimization of AKA message structure
  The PE3WAKA example protocols were designed to be examples and illustrate the concepts. The protocol(s) can be simplified further with respect to message elements. This will not necessarily lead to new results, but it may make the protocols more practical.
- Radio Access and Subscriber Privacy
  Modern radio systems are sophisticated and complex. The way that radio channels are allocated and the nature of the allocated channels are very different from the relatively simple schemes found in the 2G and indeed also in the 3G system. With OFDM/OFDMA methods in a broadband system one may permanently assign a narrow-band low latency control channel for each terminal. The availability of an always-on control channel makes it possible to provide very fast dynamic setup of variable bandwidth, variable quality (different modulation, etc.) user plane channels.
- Key Derivation and Key Hierarchies
  The issue of key hierarchy is tightly associated with the need for session keys. We have seen that in the LTE system the control plane and the user plane are separated. It may also be useful to distinguish clearly between uplink and downlink data streams. In LTE it was found that it was necessary, due to radio resource control reasons, to terminate some of the

security in the access point (eNB). This of course directly influenced the key hierarchy.

- Protocol Verification

  There is still ample room for improvements in the area of security protocol verification and it is to be hoped that the fine work done by the AVISPA project is followed up in future projects.

  Particularly the scope of the verification should be extended to permit verification of identity privacy and to allow direct support for multi-agent protocols. Similarly, it would be beneficial to have modeling language support for sequential statement execution where that best mirrors the actual protocol operations.

9.2 Final Note

This book has provided background information on the access security as found in the 3GPP systems, including the LTE system. It was demonstrated that backwards compatibility and general system complexity sometimes have led to both awkward and inelegant solutions. Despite this, the access security provided by these public systems is largely adequate and comprehensive for most use-cases. One should of course note that these systems do not provide end-to-end security, but then that was never in the scope of access security.

The research oriented part of the book explored what we can achieve and what we might require from a future cellular access security architecture. To this end we identified a set of guidelines for the design of a sound access security protocol.

We paid explicit attention to two of the areas where the 3GPP architecture(s) come up short, namely with respect to the outdated practice of delegated authentication control and with respect to providing credible personal privacy for the subscriber (identity confidentiality and location confidentiality mostly). In particular, we saw that it was fully possible to design an online authentication and key agreement that still provided the subscribers with reasonable personal privacy. Thus, there are solutions for the shortcomings of the 3GPP architecture(s). We also investigated the the topic of formal verification and concluded that while it is prudent that one go through the process of formal verification of security protocols, the most value is in the design process and not so much in the area of obtaining proofs of correctness.

Bibliography

[1] 3GPP TR 21.905. 3rd Generation Partnership Project; Technical Specification Group Services and System Aspects; Vocabulary for 3GPP Specifications. 3GPP, Sophia Antipolis, Valbonne, France, 09 2005.

[2] 3GPP TS 36.323. 3rd Generation Partnership Project; Technical Specification Group Radio Access Network; Evolved Universal Terrestrial Radio Access (E-UTRA); Packet Data Convergence Protocol (PDCP) Specification (Release 8). 3GPP, Sophia Antipolis, Valbonne, France, 03 2009.

[3] Nordic PTTs. NMT Doc 450-1: System Description. Technical Report NMT Doc 450-1, Nordic PTTs, 03 1999.

[4] Veastad, J.R., "Det er til deg ..." Historien om den første mobiltelefonen i Norge. Technical Report, Telenor Mobil, 2001.

[5] ETSI SMG. GSM TS 03.20 v6.0.1; Security Related Network Functions. ETSI, Sophia Antipolis, Valbonne, France, 07 1999.

[6] Køien, G.M. and Oleshchuk, V.A., Privacy-Preserving Spatially Aware Authentication Protocols; Analysis and Solution. In S.J. Knapskog (Ed.), *Proceedings of NORDSEC 2003*, Gjøvik, Norway, 10 2003, NORDSEC, NTNU, Norway, pp. 161–174, 2003.

[7] Køien, G.M., Overview of UMTS security for Release 99. *Telektronikk* **96**(1), pp. 102–107, 2000.

[8] Køien, G.M. and Oleshchuk, V.A., Location Privacy for Cellular Systems; Analysis and Solution. In G. Danezis and D. Martin (Eds.), *Proceedings of Privacy Enhancing Technologies workshop (PET 2005)*, Lecture Notes in Computer Science, Vol. 3856, Cavtat, Croatia, 2005. Springer, 2005, pp. 40–58.

[9] Warren, S.D. and Brandeis, L.D., The Right to Privacy. *Harward Law Review* **IV**(5), 1890.

[10] Chaum, D., Untraceable Electronic Mail, Return Addresses, and Digital Pseudonyms. *Communications of the ACM* **24**(2), 1981.

[11] Federrath, H., Jerichow, A. and Pfitzmann, A., MIXes in Mobile Communication Systems: Location Management with Privacy. In *Proceedings of the First Intern. Workshop on Information Hiding*, Cambridge, UK, Lecture Notes in Computer Science, Vol. 1174, Springer, pp. 121–135, 1996.

[12] Samfat, D., Molva, R. and Asokan, N., Untraceability in Mobile Networks. In *The First International Conference on Mobile Computing and Networking (ACM MOBICOM 95)*, Berkely, California, USA. ACM Press, pp. 26–36, 1995.

[13] Ateniese, G., Herzberg, A., Krawczyk, H. and Tsudik, G., Untraceable Mobility or How to Travel Incognito, Volume 31. Elsevier, 1999.

[14] Go, J. and Kim, K., Wireless Authentication Protocol Preserving User Anonymity. In *The 2001 Symposium on Cryptography and Information Security (SCIS 2001)*, Oiso, Japan. IEICE, pp. 26–36, 2001.

[15] Asokan, N., Anonymity in a Mobile Computing Environment. In *Proceedings of IEEE Workshop on Mobile Computing Systems and Applications*, Santa Cruz, CA, USA. IEEE Press, pp. 200–204, 1994.

[16] ETSI SMG. GSM TS 03.03 v6.6.0; Numbering, Addressing and Identification. ETSI, Sophia Antipolis, Valbonne, France, 06 2000.

[17] 3GPP TS 23.003. 3rd Generation Partnership Project; Technical Specification Group Core Network and Terminals; Numbering, Addressing and Identification (Release 7). 3GPP, Sophia Antipolis, Valbonne, France, 06 2007.

[18] ITU-T. Rec. E.212; The International Identification Plan for Mobile Terminals and Mobile Users. International Telecommunication Union, Geneva, Switzerland, 05 2004.

[19] ITU-T. Rec. E.164; The International Public Telecommunication Numbering Plan. International Telecommunication Union, Geneva, Switzerland, 02 2005.

[20] ETSI SMG. GSM TS 02.09 v6.0.1; Security aspects. ETSI, Sophia Antipolis, Valbonne, France, 05 1999.

[21] Krawczyk, H., How to Predict Congruential Generators. *Journal of Algorithms* **13**(4), 1992.

[22] Menezes, A.J., van Oorschot, P.C. and Vanstone, S.A., *Handbook of Applied Cryptography*, Revised Reprint with Updates, 5th printing. CRC Press, Boca Raton, FL, 2001.

[23] Ferguson, N. and Schneier, B., *Practical Cryptography*. Wiley, Indianapolis, IN, 2003.

[24] Brookson, C., Can you clone a GSM Smart Card (SIM)? Available at http://www.brookson.com/gsm/contents.htm, 2002.

[25] Briceno, M., Goldberg, I. and Wagner, D., GSM Cloning. Available at http://www.isaac.cs.berkeley.edu/isaac/gsm-faq.html, 1998.

[26] 3GPP TS 55.205. 3rd Generation Partnership Project; Technical Specification Group Services and System Aspects; Specification of the GSM-MILENAGE Algorithms: An example algorithm set for the GSM Authentication and Key Generation functions A3 and A8 (Release 6). 3GPP, Sophia Antipolis, Valbonne, France, 06 2006.

[27] Biryukov, A., Shamir, A. and Wagner, D., Real Time Cryptanalysis of A5/1 on a PC. In *Proceedings of Fast Software Encryption, 7th International Workshop, FSE 2000*, New York, NY, USA, April 10–12, 2000, Lecture Notes in Computer Science, Vol. 1978, Springer, pp. 1–18, 2001.

[28] Barkan, E., Biham, E. and Keller, N., Instant Ciphertext-only Cryptanalysis of GSM Encrypted Communication. In *Proceedings of CRYPTO 2003, 23rd Annual International Cryptology Conference*, Santa Barbara, CA, USA, August 17–21, 2003, Lecture Notes in Computer Science, Vol. 2729, Springer, 2003.

[29] ETSI SMG. GSM TS 02.17 v5.0.1; Subscriber Identity Modules (SIM); Functional Characteristics. ETSI, Sophia Antipolis, Valbonne, France, 04 1997.

[30] ETSI SMG. GSM TS 11.11 v5.5.1; Mobile Equipment (SIM–ME) Interface. ETSI, Sophia Antipolis, Valbonne, France, 10 1997.

[31] ISO/IEC. 7810: Identification Cards – Physical Characteristics. International Organization for Standardization (ISO), Geneva, Switzerland, 1995.

[32] ETSI SMG. GSM TS 04.08 v5.8.0; Mobile Radio Interface Layer 3 Specification. ETSI, Sophia Antipolis, Valbonne, France, 01 1998.

[33] Shannon, C.E., Communication Theory of Secrecy Systems. *Bell System Technical Journal* 1948. Also in *Claude Elwood Shannon – Collected Papers*, N.J.A. Sloane and A.D. Wyner (Eds.). IEEE Press, 1993.

[34] 3GPP TS 33.102. 3rd Generation Partnership Project; Technical Specification Group Services and System Aspects; 3G Security; Security architecture (Release 8). 3GPP, Sophia Antipolis, Valbonne, France, 03 2009.

[35] 3GPP TS 31.101. 3rd Generation Partnership Project; Technical Specification Group Core Network and Terminals; UICC-terminal Interface; Physical and logical Characteristics (Release 8). 3GPP, Sophia Antipolis, Valbonne, France, 01 2009.

[36] 3GPP TS 31.102. 3rd Generation Partnership Project; Technical Specification Group Core Network and Terminals; Characteristics of the Universal Subscriber Identity Module (USIM) Application (Release 8). 3GPP, Sophia Antipolis, Valbonne, France, 03 2009.

[37] 3GPP TS 33.120. 3rd Generation Partnership Project; Technical Specification Group Services and System Aspects; 3G Security; Security Principles and Objectives (Release 4). 3GPP, Sophia Antipolis, Valbonne, France, 03 2001.

[38] 3GPP TS 21.133. 3rd Generation Partnership Project; Technical Specification Group Services and System Aspects; 3G Security; Security Threats and Requirements (Release 4). 3GPP, Sophia Antipolis, Valbonne, France, 12 2001.

[39] Køien, G.M., An Introduction to Access Security in UMTS. *IEEE Wireless Communications Magazine* **11**(1), 8–18, 2004.

[40] Boman, K., Horn, G., Howard, P. and Niemi, V., UMTS Security. *Electronic & Communications Engineering Journal*, 191–204, 2002.

[41] Niemi, V. and Nyberg, K., *UMTS Security*. Wiley, England, 2003.

[42] 3GPP TS 33.105. 3rd Generation Partnership Project; Technical Specification Group Services and System Aspects; 3G Security; Cryptographic Algorithm Requirements (Release 6). 3GPP, Sophia Antipolis, Valbonne, France, 06 2004.

[43] 3GPP TS 35.205. 3rd Generation Partnership Project; Technical Specification Group Services and System Aspects; 3G Security; Specification of the MILENAGE Algorithm Set: An example algorithm set for the 3GPP authentication and key generation functions f1, f1*, f2, f3, f4, f5 and f5*; Document 1: General (Release 6). 3GPP, Sophia Antipolis, Valbonne, France, 12 2004.

[44] 3GPP TS 35.206. 3rd Generation Partnership Project; Technical Specification Group Services and System Aspects; 3G Security; Specification of the MILENAGE Algorithm Set: An example algorithm set for the 3GPP authentication and key generation functions f1, f1*, f2, f3, f4, f5 and f5*; Document 2: Algorithm Specification (Release 6). 3GPP, Sophia Antipolis, Valbonne, France, 12 2004.

[45] 3GPP TS 35.207. 3rd Generation Partnership Project; Technical Specification Group Services and System Aspects; 3G Security; Specification of the MILENAGE Algorithm Set: An example algorithm set for the 3GPP authentication and key generation functions f1, f1*, f2, f3, f4, f5 and f5*; Document 3: Implementors' Test Data (Release 6). 3GPP, Sophia Antipolis, Valbonne, France, 12 2004.

[46] 3GPP TS 35.208. 3rd Generation Partnership Project; Technical Specification Group Services and System Aspects; 3G Security; Specification of the MILENAGE Algorithm

Set: An example algorithm set for the 3GPP authentication and key generation functions f1, f1*, f2, f3, f4, f5 and f5*; Document 4: Design Conformance Test Data (Release 6). 3GPP, Sophia Antipolis, Valbonne, France, 12 2004.

[47] 3GPP TR 35.909. 3rd Generation Partnership Project; Technical Specification Group Services and System Aspects; 3G Security; Specification of the MILENAGE Algorithm Set: An example algorithm set for the 3GPP authentication and key generation functions f1, f1*, f2, f3, f4, f5 and f5*; Document 5: Summary and results of design and evaluation (Release 6). 3GPP, Sophia Antipolis, Valbonne, France, 12 2004.

[48] NIST. Federal Information Processing Standards Publication 197; ADVANCED ENCRYPTION STANDARD (AES). National Institute of Standards and Technology (NIST), 11 2001.

[49] 3GPP2 S.S0055. 3rd Generation Partnership Project 2; Enhanced Cryptographic Algorithms (Version 1.0). 3GPP2, 01 2002.

[50] 3GPP TR 33.909. 3rd Generation Partnership Project; Technical Specification Group Services and System Aspects; 3G Security; Report on the Design and Evaluation of the MILENAGE Algorithm Set; Deliverable 5: An Example Algorithm for the 3GPP Authentication and Key Generation Functions (Release 4). 3GPP, Sophia Antipolis, Valbonne, France, 06 2001.

[51] 3GPP TS 35.202. 3rd Generation Partnership Project; Technical Specification Group Services and System Aspects; 3G Security; Specification of the 3GPP Confidentiality and Integrity Algorithms; Document 2: KASUMI Specification (Release 6). 3GPP, Sophia Antipolis, Valbonne, France, 12 2004.

[52] 3GPP TR 33.908. 3rd Generation Partnership Project; Technical Specification Group Services and System Aspects; 3G Security; General Report on the Design, Speification and Evaluation of 3GPP Standard Confidentiality and Integrity Algorithms (Release 4). 3GPP, Sophia Antipolis, Valbonne, France, 09 2001.

[53] 3GPP TS 35.215. 3rd Generation Partnership Project; Technical Specification Group Services and System Aspects; Specification of the 3GPP Confidentiality and Integrity Algorithms UEA2 & UIA2; Document 1: UEA2 and UIA2 specifications (Release 7). 3GPP, Sophia Antipolis, Valbonne, France, 06 2006. *Subject to licensing*.

[54] 3GPP TS 35.216. 3rd Generation Partnership Project; Technical Specification Group Services and System Aspects; Specification of the 3GPP Confidentiality and Integrity Algorithms UEA2 & UIA2; Document 2: SNOW 3G specification (Release 7). 3GPP, Sophia Antipolis, Valbonne, France, 06 2006. *Subject to licensing*.

[55] 3GPP TS 35.217. 3rd Generation Partnership Project; Technical Specification Group Services and System Aspects; Specification of the 3GPP Confidentiality and Integrity Algorithms UEA2 & UIA2; Document 3: Implementors' test data (Release 7). 3GPP, Sophia Antipolis, Valbonne, France, 06 2006. *Subject to licensing*.

[56] 3GPP TS 35.218. 3rd Generation Partnership Project; Technical Specification Group Services and System Aspects; Specification of the 3GPP Confidentiality and Integrity Algorithms UEA2 & UIA2; Document 4: Design conformance test data (Release 7). 3GPP, Sophia Antipolis, Valbonne, France, 06 2006. *Subject to licensing*.

[57] 3GPP TR 35.919. 3rd Generation Partnership Project; Technical Specification Group Services and System Aspects; 3G Security; Specification of the 3GPP Confidentiality and Integrity Algorithms UEA2 & UIA2: Document 5: Design and Evaluation Report (Release 7). 3GPP, Sophia Antipolis, Valbonne, France, 03 2006. *Subject to licensing*.

[58] 3GPP TS 29.002. 3rd Generation Partnership Project; Technical Specification Group Core Network and Terminals; Mobile Application Part (MAP) specification;. 3GPP, Sophia Antipolis, Valbonne, France, 03 2006.

[59] 3GPP TS 29.060. 3rd Generation Partnership Project; Technical Specification Group Core Networks; Feasibility Study on SS7 signalling transport in the core network with SCCP-User Adaptation (SUA) layer. 3GPP, Sophia Antipolis, Valbonne, France, 12 2004.

[60] Loughney, J., Tuexen, M. and Pastor-Balbas, J., RFC 3788: Security Considerations for Signaling Transport (SIGTRAN) Protocols. The Internet Engineering Task Force (IETF), http://www.ietf.org/, 06 2004.

[61] 3GPP TS 24.008. 3rd Generation Partnership Project; Technical Specification Group Core Network and Terminals; Mobile radio interface Layer 3 specification; Core network protocols; Stage 3(Release 7). 3GPP, Sophia Antipolis, Valbonne, France, 09 2005.

[62] Køien, G.M., A Validation Model of the UMTS Authentication and Key Agreement Protocol. Research note 59/2002, Telenor R&D, 2002. Open.

[63] 3GPP TR 33.902. 3rd Generation Partnership Project; Technical Specification Group Services and System Aspects; 3G Security; Formal Analysis of the 3G Authentication Protocol (Release 4). 3GPP, Sophia Antipolis, Valbonne, France, 09 2001.

[64] Lamport, L., *Specifying Systems: The TLA+ Language and Tools for Hardware and Software Engineers*. Addison-Wesley Professional, 07 2002.

[65] Wedel, G. and Kessler, V., Formal Semantics for Authentication Logics. In *Proceedings of 4th European Symposium on Research in Computer Security (ESORICS 96)*, Rome, Italy, September 25–27, 1996, Lecture Notes in Computer Science, Vol. 1146, Springer, pp. 218–241, 1996.

[66] Burrows, M., Abadi, M., and Needham, R., A Logic of Authentication. Research Report 39, DEC Systems Research Center, Palo Alto, California, USA, 2 1990.

[67] Holzmann, G.J., *The Spin Model Checker: Primer and Reference Manual*. Addison-Wesley Professional, 2003.

[68] AVISPA Project. Automated Validation of Internet Security Protocols and Applications (AVISPA); Deliverable D2.1: The High Level Protocol Specification Language. AVISPA IST-2001-39252, http://www.avispa-project.org, 08 2003.

[69] 3GPP TS 33.210. 3rd Generation Partnership Project; Technical Specification Group Services and System Aspects; 3G Security; Network Domain Security; IP Network Layer Security (Release 6). 3GPP, Sophia Antipolis, Valbonne, France, 06 2004.

[70] ETSI SMG. GSM TS 09.02 v6.2.0; Mobile Application Part (MAP) specification. ETSI, Sophia Antipolis, Valbonne, France, 11 1998.

[71] Køien, G.M., UMTS Network Domain Security. In *Proceedings of EURESCOM SUMMIT 2002*, Heidelberg, Germany. EURESCOM, VDE Verlag, pp. 185–194, 2002.

[72] 3GPP TS 33.310. 3rd Generation Partnership Project; Technical Specification Group Services and System Aspects; Network Domain Security (NDS); Authentication Framework (AF) (Release 6). 3GPP, Sophia Antipolis, Valbonne, France, 09 2004.

[73] Rose, G. and Køien, G.M., Access Security in CDMA200, Including a Comparison with UMTS Access Security. *IEEE Wireless Communications magazine* 11(1), 19–25, 2004.

[74] 3GPP2 S.S0055. 3rd Generation Partnership Project 2; Common Security Algorithms (Version 1.0). 3GPP2, 12 2002.

[75] Kent, S. and Seo, K., RFC 4301: Security Architecture for the Internet Protocol. The Internet Engineering Task Force (IETF), http://www.ietf.org/, 12 2005.

[76] Anderson, R., Why Cryptosystems Fail. *Communications of the ACM* **37**(11), 32–40, 1994.

[77] 3GPP TR 36.913. 3rd Generation Partnership Project; Technical Specification Group Radio Access Network; Requirements for Further Advancements for E-UTRA (LTE-Advanced) (Release 8). 3GPP, Sophia Antipolis, Valbonne, France, 06 2008.

[78] 3GPP TS 33.401. 3rd Generation Partnership Project; Technical Specification Group Services and System Aspects; 3GPP System Architecture Evolution (SAE): Security Architecture (Release 8). 3GPP, Sophia Antipolis, Valbonne, France, 03 2009.

[79] 3GPP TS 33.402. 3rd Generation Partnership Project; Technical Specification Group Services and System Aspects; 3GPP System Architecture Evolution (SAE): Security aspects of non-3GPP accesses (Release 8). 3GPP, Sophia Antipolis, Valbonne, France, 03 2009.

[80] 3GPP TS 23.003. 3rd Generation Partnership Project; Technical Specification Group Core Network and Terminals; Network architecture (Release 8). 3GPP, Sophia Antipolis, Valbonne, France, 12 2008.

[81] 3GPP TS 23.401. 3rd Generation Partnership Project; General Packet Radio Service (GPRS) enhancements for Evolved Universal Terrestrial Radio Access Network (E-UTRAN) access (Release 9) . 3GPP, Sophia Antipolis, Valbonne, France, 03 2009.

[82] 3GPP TS 36.300. 3rd Generation Partnership Project; Technical Specification Group Radio Access Network; Evolved Universal Terrestrial Radio Access (E-UTRA) and Evolved Universal Terrestrial Radio Access Network (E-UTRAN); Overall description; Stage 2 (Release 8). 3GPP, Sophia Antipolis, Valbonne, France, 03 2009.

[83] 3GPP TS 36.401. 3rd Generation Partnership Project; Technical Specification Group Radio Access Network; Evolved Universal Terrestrial Radio Access Network (E-UTRAN); Architecture description (Release 8). 3GPP, Sophia Antipolis, Valbonne, France, 03 2009.

[84] 3GPP TS 36.331. 3rd Generation Partnership Project; Technical Specification Group Radio Access Network; Evolved Universal Terrestrial Radio Access (E-UTRA); Radio Resource Control (RRC); Protocol specification (Release 8). 3GPP, Sophia Antipolis, Valbonne, France, 03 2009.

[85] 3GPP TS 24.301. 3rd Generation Partnership Project; Technical Specification Group Core Network and Terminals; Non-Access-Stratum (NAS) protocol for Evolved Packet System (EPS); Stage 3 (Release 8). 3GPP, Sophia Antipolis, Valbonne, France, 09 2009.

[86] 3GPP TS 29.060. 3rd Generation Partnership Project; Technical Specification Group Core Network and Terminals; General Packet Radio Service (GPRS); GPRS Tunnelling Protocol (GTP) across the Gn and Gp interface (Release 6). 3GPP, Sophia Antipolis, Valbonne, France, 03 2006.

[87] 3GPP TS 29.281. 3rd Generation Partnership Project; Technical Specification Group Core Network and Terminals; General Packet Radio System (GPRS) Tunnelling Protocol User Plane (GTPv1-U) (Release 8). 3GPP, Sophia Antipolis, Valbonne, France, 03 2009.

[88] Calhoun, P., Loughney, J., Guttman, E., Zorn, G. and Arkko, J., RFC 3588: Diameter Base Protocol. The Internet Engineering Task Force (IETF), http://www.ietf.org/, 09 2003.

[89] Calhoun, P., Zorn, G., Spence, D. and Mitton, D., RFC 4005: Diameter Network Access Server Application. The Internet Engineering Task Force (IETF), http://www.ietf.org/, 08 2005.

[90] 3GPP TS 29.272. 3rd Generation Partnership Project; Technical Specification Group Core Network and Terminals; Evolved Packet System (EPS); Mobility Management Entity (MME) and Serving GPRS Support Node (SGSN) related interfaces based on Diameter protocol (Release 8). 3GPP, Sophia Antipolis, Valbonne, France, 03 2009.

[91] 3GPP TS 29.274. 3rd Generation Partnership Project; Technical Specification Group Core Network and Terminals; 3GPP Evolved Packet System (EPS); Evolved General Packet Radio Service (GPRS) Tunnelling Protocol for Control plane (GTPv2-C); Stage 3 (Release 8). 3GPP, Sophia Antipolis, Valbonne, France, 03 2009.

[92] R. Stewart, R. (Ed.), RFC 4960: Stream Control Transmission Protocol. The Internet Engineering Task Force (IETF), http://www.ietf.org/, 09 2007.

[93] 3GPP TS 36.423. 3rd Generation Partnership Project; Technical Specification Group Radio Access Network; Evolved Universal Terrestrial Radio Access Network (E-UTRAN); X2 application protocol (X2AP) (Release 8). 3GPP, Sophia Antipolis, Valbonne, France, 03 2009.

[94] 3GPP TR 33.820. 3rd Generation Partnership Project; Technical Specification Group Service and System Aspects; Security of H(e)NB (Release 8). 3GPP, Sophia Antipolis, Valbonne, France, 03 2009. (Internal report/only for internal discussion.)

[95] 3GPP TS 33.220. 3rd Generation Partnership Project; Technical Specification Group Services and System Aspects; Generic Authentication Architecture (GAA); Generic bootstrapping architecture (Release 7). 3GPP, Sophia Antipolis, Valbonne, France, 06 2005.

[96] Krawczyk, H., Bellare, M. and Canetti, R., RFC 2104: HMAC: Keyed-Hashing for Message Authentication. The Internet Engineering Task Force (IETF), http://www.ietf.org/, 02 1997.

[97] NIST. Federal Information Processing Standards Publication 180-2; SECURE HASH STANDARD. National Institute of Standards and Technology (NIST), 08 2002.

[98] NIST. NIST Special Publication 800-38A; 2001 Edition; Recommendation for Block Cipher Modes of Operation. National Institute of Standards and Technology (NIST), 12 2001.

[99] NIST. NIST Special Publication 800-38B; 2001 Edition; Recommendation for Block Cipher Modes of Operation: The CMAC Mode for Authentication. National Institute of Standards and Technology (NIST), 05 2005.

[100] Dolev, D. and Yao, A., On the Security of Public-Key Protocols. *IEEE Transactions on Information Theory* **29**(2), 198–208, 1983.

[101] Kömmerling, O. and Kuhn, M.G., Design Principles for Tamper-Resistant Smartcard Processors. In *USENIX Workshop on Smartcard Technology*, Chicago, Illinois, USA. USENIX, USENIX, 1999.

[102] Benoit, O., Dabbous, N., Gauteron, L., Girard, P., Handschuh, H., Naccache, D., Socié, S. and Whelan, C., Mobile Terminal Security. Number 2004-184. IACR Cryptology ePrint Archive, http://eprint.iacr.org/2004/158, 07 2004.

[103] Bar-El, H., Choukri, H., Naccache, D., Tunstall, M. and Whelan, C., The Sorcerer's Apprentice Guide to Fault Attacks. *Proceedings of the IEEE* **94**, 370–382, 2006.

[104] Coron, J.-S., Naccache, D. and Kocher, P., Statistics and Secret Leakage. *ACM Transactions on Embedded Computing Systems (TECS)*, **3**, 492–508, 2004.

[105] ETSI SR 002 180. Requirements for Communications of Citizens with Authorities/Organizations in Case of Distress (Emergency Call Handling). ETSI, Sophia Antipolis, Valbonne, France, 12 2003.

[106] EU Council/EU Parliament. Directive of the European Parliament and of the Council on the retention of data generated or processed in connection with the provision of publicly available electronic communications services or of public communications networks and amending Directive 2002/58/EC. EU Directive, 2 2006. Data Retention Directive.

[107] 3GPP TS 33.234. 3rd Generation Partnership Project; Technical Specification Group Services and System Aspects; 3G Security; Wireless Local Area Network (WLAN) interworking security (Release 6). 3GPP, Sophia Antipolis, Valbonne, France, 06 2005.

[108] Køien, G.M. and Haslestad, T., Security Aspects of 3G-WLAN Interworking. *IEEE Communications magazine* **41**(11), 82–89, 2002.

[109] Joux, A., A One Round Protocol for Tripartite DiffieHellman. In W. Bosma (Ed.), *Proceedings ANTS-IV 2000*, Leiden, The Netherlands, Lecture Notes in Computer Science, Vol. 1838. Springer, pp. 385–394, 2000.

[110] Joux, A., A One Round Protocol for Tripartite DiffieHellman. *Journal of Cryptology* **17**(4), 263–276, 2004. (Updated and revised version of [109].)

[111] Hofheinz, D., Müller-Quade, J. and Steinwandt, R., Initiator-Resilient Universally Composable Key Exchange. In D. Gollmann and E. Snekkenes (Eds.), *Proceedings of ESORICS 2003*, Gjøvik, Norway, Lecture Notes in Computer Science, Vol. 2808. Springer, 2003.

[112] Køien, G.M., Principles for Cellular Access Security. In S. Liitmatainen and T. Virtanen (Eds.), *Proceedings of NORDSEC 2004*, Espoo, Finland. NORDSEC, HUT, Finland, pages 65–72, 2004.

[113] Køien, G.M., Privacy Enhanced Cellular Access Security. In *Proceedings of the 2005 ACM Workshop on Wireless Security*, Cologne, Germany. ACM SIGmobile, ACM Press, pp. 57–66, 2005.

[114] Køien, G.M. and Oleshchuk, V.A., Spatio-Temporal Exposure Control. In *Proceedings of IEEE PIMRC 2003*, Volume 14, Beijing, China. IEEE PIMRC, IEEE Press. pp. 2760–2764, 2003.

[115] Hansen, F.Ø. and Oleshuk, V.A., Spatial Role-Based Access Control Model for Wireless Networks. In *Proceedings of IEEE Vehicular Technology Conference VTC2003-Fall*, Vol. 3, Orlando, USA. IEEE Press, pp. 2093–2097, 2003.

[116] Gligoroski, D., Andova, S. and Knapskog, S.-J., On the Importance of the Key Separation Principle for Different Modes of Operation. In *Proceedings of the 4th Information Security Practice and Experience Conference (ISPEC 2008)*, Sydney, Australia, 21–23 April 2008.

[117] GSMA IREG Work Group. Inter-PLMN Backbone Guidelines (v.3.7). Permanent reference document IR.34, GSM Association, 04 2006. Available at: http://www.gsmworld.com/documents/ireg/ir34.pdf.

[118] ECRYPT Project. European Network of Excellence in Cryptology (ECRYPT); ECRYPT Yearly Report on Algorithms and Keysizes (2007-2008). D.SPA.28, ECRYPT IST-2002-507932, 07 2008.

[119] Walker, J., Unsafe at Any Key Size: An Analysis of the WEP Encapsulation. Technical Report 03628E, IEEE 802.11 Committee, 03 2000.

[120] Morris Dworkin, M. (Ed.), NIST Special Publication 800-38D; DRAFT (June 2007); Recommendation for Block Cipher Modes of Operation: Galois/Counter Mode (GCM) for Confidentiality and Authentication. National Institute of Standards and Technology (NIST), 06 2007.

[121] Boyd, C. and Mathuria, A., *Protocols for Authentication and Key Establishment*. Information Security and Cryptography; Texts and Monographs. Springer, Germany, 2003.

[122] Carlsen, U., Cryptographic Protocol Flaws. In *Proceedings of the 1994 IEEE Computer Security Foundations Workshop VII*, IEEE Computer Society Press, pp. 192–200, 1994.

[123] Gritzalis, S.and Spinellis, D., Cryptographic Protocols over Open Distributed Systems: A Taxonomy of Flaws and related Protocol Analysis Tools. In P. Daniel (Ed.), *16th International Conference on Computer Safety, Reliability and Security: SAFECOMP '97*, York, UK. Springer Verlag, pages 123–137, 1997.

[124] CERT. TCP SYN Flooding and IP Spoofing Attacks. CERT Advisory CA-1996-21 1996, available at http://www.cert.org/advisories/CA-1996-21.html), 09 1996.

[125] Clark, J. and Jacob, J., A Survey of Authentication Protocol Literature: Version 1.0. Technical Report, University of York, England, http://www-users.cs.york.ac.uk/ jac/papers/drareviewps.ps, 11 1997.

[126] Electronic Frontier Foundation. *Cracking DES; Secrets of Encryption Research, Wiretap Politics & Chip Design*. O'Reilly Press, 1998.

[127] Syverson, P., Meadows, C. and Cervesato, I., Dolev–Yao Is No Better Than Machiavelli. In P. Degano (Ed.), *First Workshop on Issues in the Theory of Security – WITS'00*, Geneva, Switzerland, 2000.

[128] Blaze, M., Diffie, W., Rivest, R., Schneier, B., Shimomura, T., Thompson, E. and Wiener, M., Minimimal Key Lengths for Symmetric Ciphers to Provide Adequate Commercial Security. Report of ad hoc panel of cryptographers and computer scientists, http://theory.lcs.mit.edu/~rivest/bsa-final-report.pdf, 01 1996.

[129] ITU-T. Rec. E.214; Structure of the land mobile global title for the signalling connection control part (SCCP). International Telecommunication Union, Geneva, Switzerland, 02 2005.

[130] Krawczyk, H., SIGMA: The 'SIGn-and-MAc' Approach to Authenticated Diffie-Hellman and Its Use in the IKE Protocols. In *Proceedings of CRYPTO 2005, 23rd Annual International Cryptology Conference, Santa Barbara, California, USA, August 17–21, 2003*, Lecture Notes in Computer Science, Vol. 2729. Springer, pp. 400–425, 2005.

[131] Kesdogan, D., Federrath, H., Jerichow, A. and Pfitzmann, A., Location Management Strategies Increasing Privacy in Mobile Communication. In *Proceedings of IFIP SEC 1996*, pp. 39–48, 1996.

[132] Atallah, M.J. and Du, W., Secure Multy-Party Computational Geometry. In *Proceedings of WADS2001: 7th International Workshop on Algorithms and Data Structures*, Rhode

Island, USA. Lecture Notes in Computer Science, Vol. 2125. Springer, pp. 165–179, 2001.

[133] Du, W. and Atallah, M.J., Secure Multy-Party Computation Problems and Their Applications: A Review and Open Problems. In *Proceedings of the 2001 workshop on New Security Paradigms (NSPW'01)*, New Mexico, USA. ACM Press, pp. 13–22, 2001.

[134] Goldreich, O., Micali, S. and Wigderson, A., How to Play Any Mental Game. In A.V. Aho (Ed.), *Proceedings of the 19th Annual ACM Symposium on Theory of Computing*, New York, USA. ACM Press, pp. 218–229, 1987.

[135] Du, W. and Zhan, Z., A Practical Approach to Solve Secure Multi-Party Computational Problems. In *Proceedings of the 2002 Workshop on New Security Paradigms*, Virginia Beach, Virginia, USA. ACM Press, pp. 127–135, 2002.

[136] Huang, L., Yamane, H. and Matsuura Kaoru Sezaki, K., Towards Modelling Wireless Location Privacy. In G. Danezis and D. Martin (Eds.), *Proceedings of Privacy Enhancing Technologies Workshop (PET 2005)*, Cavtat, Croatia, Lecture Notes in Computer Science, Vol. 3856. Springer, pp. 59–77, 2005.

[137] Harkins, D. and Carrel, D., RFC 2409: The Internet Key Exchange (IKE). The Internet Engineering Task Force (IETF), http://www.ietf.org/, 11 1998.

[138] Kaufman, C. (Ed.), RFC 4306: Internet Key Exchange (IKEv2) Protocol. The Internet Engineering Task Force (IETF), http://www.ietf.org/, 12 2005.

[139] Abadi, M. and Needham, R., Prudent Engineering Practice for Cryptographic Protocols. Research Report 125, DEC Systems Research Center, Palo Alto, California, USA, 6 1994.

[140] Anderson, R. and Needham, R., Robustness Principles for Public Key Protocols. In D. Coppersmith, editor, *Advances in Cryptology – CRYPTO '95: 15th Annual International Cryptology Conference, Santa Barbara, California, USA, August 1995*, Lecture Notes in Computer Science, Vol. 963. Springer, pp. 236–247, 1995.

[141] Bird, R., Gopal, I., Herzberg, A., Janson, P.A., Kutten, S., Molva, R. and Yung, M., Systematic Design of a Family of Attack-Resistant Authentication Protocols. *IEEE Journal on Selected Areas in Communication* 11(5), 679–693, 1993.

[142] Aiello, W., Bellovin, S.M., Blaze, M., Canetti, R., Ioannidis, J., Keromytis, A.D. and Reingold, O., Efficient, DoS-Resistant, Secure Keu Exchange for Internet Protocols. In *Proceedings of the 9th ACM conference on Computer and Communications Security (CCS'02)*, Washington, DC, USA. ACM Press, pp. 48–58, 2002.

[143] Canetti, R., Meadows, C. and Syverson, P., Environmental Requirements for Authentication. In *Software Security – Theories and Systems; Mext-NSF-JSPS International Symposium, ISSS 2002*, Tokyo, Japan, November 8–10, 2002, Revised Papers, Lecture Notes in Computer Science, Vol. 2609. Springer, pp. 339–355, 2003.

[144] 3GPP TR 33.978. 3rd Generation Partnership Project; Technical Specification Group Services and System Aspects; Security aspects of early IP Multimedia Subsystem (IMS) (Release 6). 3GPP, Sophia Antipolis, Valbonne, France, 06 2005.

[145] Syverson, P., Limitations on Design Principles for Public Key Protocols. In *Proceedings of IEEE Symposium on Security and Privacy*. IEEE Press, pp. 62–72, 1996.

[146] Milne, A.A., *When We Were Very Young*, original edition. Puffin Books (Penguin Group), 1928.

[147] Needham, R.M. and Schroeder, M.D., Using Encryption for Authentication in Large Networks of Computers. *Communications of the ACM* 21(12), 993–999, 1978.

[148] Lowe, G., An Attack on the Needham–Schroeder Public-Key Authentication protocol. In R.S. Bird (Ed.), *Information Processing Letters*, No. 56. Elsevier Science, pp. 131–133, 1995.

[149] Lowe, G., Breaking and Fixing the Needham-Schroeder Public-Key Protocol Using FDR. In *Proceedings of Second International Workshop, TACAS '96*, Passau, Germany, March 27-29, 1996, Lecture Notes in Computer Science, Vol. 1055. Springer, pp. 147–166, 1996.

[150] Lampson, B., Abadi, M., Burrows, M. and Wobber, E., Authentication in Distributed Systems: Theory and Practice. *ACM Transactions on Computer Systems (TOCS)* 10(4), 265–310, 1992. (A preliminary version was published in *ACM SIGOPS Operating Systems Review* 25(5), October, 1991.)

[151] Bellare, M. and Rogaway, P., Provably Secure Session Key Distribution: The Three Party Case. In *Proceedings of the Twenty-Seventh Annual ACM Symposium on Theory of Computing (STOC '95)*, Las Vegas, Nevada, USA. ACM Press, pp. 57–66, 1995.

[152] Gong, L., Lower Bounds on Messages and Rounds for Network Authentication Protocols. In *Proceedings of the 1st ACM Conference on Computer and Communications Security (CCS '93)*, Fairfax, Virginia, USA. ACM Press, 1993, pp. 26–37 .

[153] Lowe, G., A Hierarchy of Authentication Specifications. In *Proceedings of the 10th Computer Security Foundations Workshop (CSFW '97)*, Rockport, Massachusetts, USA. IEEE Computer Society, pp. 31–43, 1997.

[154] ITU-T. Rec. X.509; Information Technology – Open Systems Interconnection – The Directory: Public-key and attribute certificate frameworks. International Telecommunication Union, Geneva, Switzerland, 08 2005.

[155] ISO/IEC. 9798-1: Information Technology – Security Techniques – Entity Authentication – Part 1: General. International Organization for Standardization (ISO), Geneva, Switzerland, 2nd edition, 1997.

[156] ISO/IEC. 9798-2: Information Technology – Security Techniques – Entity Authentication – Part 2: Mechanisms Using Symmetric Encipherment Algorithms. International Organization for Standardization (ISO), Geneva, Switzerland, 2nd edition, 1999.

[157] ISO/IEC. 9798-3: Information Technology – Security Techniques – Entity Authentication – Part 3: Mechanisms Using Digital Signature Techniques. International Organization for Standardization (ISO), Geneva, Switzerland, 2nd edition, 1998.

[158] ISO/IEC. 9798-4: Information Technology – Security Techniques – Entity Authentication – Part 4: Mechanisms Using a Cryptographic Check Function. International Organization for Standardization (ISO), Geneva, Switzerland, 2nd edition, 12 1999.

[159] Abadi, M., Two Facets of Authentication. In *Proceedings of the 11th IEEE Computer Security Foundations Workshop*, Rockport, MA, USA. IEEE Press, pp. 27–32, 1998.

[160] Park, D., Boyd, C., Lee, B. and Kim, H., Responsibility and Credit: New Members of the Authentication Family? In *Joint Workshop on Foundations of Computer Security and Automated Reasoning for Security Protocols Analysis (FCS-ARSPA '06)*, Seattle, USA, pp. 195–210, 2006. Informal proceedings; published at http://www.easychair.org/FLoC-06/FCS-ARSPA.html.

[161] ITU-T. Rec. X.800; Security architecture for Open Systems Interconnection for CCITT Applications. International Telecommunication Union, Geneva, Switzerland, 03 1991. (ITU-T Rec. X.800 is technically aligned with ISO 7498-2.)

[162] Diffie, W. and Hellman, M.E., New Directions in Cryptography. *IEEE Transactions on Information Theory* **22**(6), 644–654, 1976.

[163] Lauter, K., The Advantages of Elliptic Curve Cryptography for Wireless Security. *IEEE Wireless Communications Magazine* **11**(1), 62–67, 2004.

[164] Diffie, W., van Oorschot, P.C. and Wiener, M.J., Authentication and Authenticated Key Exchanges. *Designs, Codes and Cryptography* **2**(2), 107–125, 1992.

[165] 3GPP TS 23.108. 3rd Generation Partnership Project; Technical Specification Group Core Network (CN); Mobile radio interface layer 3 specification, Core network protocols; Stage 2 (Release 7) . 3GPP, Sophia Antipolis, Valbonne, France, 06 2007.

[166] Køien, G.M., Entity Authentication and Personal Privacy in Future Cellular Systems. PhD Thesis, Aalborg University, April 2008.

[167] Gollmann, D., Analysing Security Protocols. In A.E. Abdalla, P. Ryan and S. Schneider (Eds.), *Formal Aspects of Security*, FASec, London, UK, 12 2002. Lecture Notes in Computer Science, Vol. 2629. Springer, pp. 71–80, 2002.

[168] Abadi, M. and Gordon, A.D., A Calculus for Cryptographic Protocols; The Spi Calculus. In *Proceedings of the 4th ACM conference on Computer and Communications Security 1997*, Zurich, Switzerland, April 01–04. ACM Press, pp. 36–47, 1997.

[169] Abadi, M. and Gordon, A.D., A Calculus for Cryptographic Protocols; The Spi Calculus. Research Report 149, DEC Systems Research Center, Palo Alto, California, USA, 1 1998.

[170] Abadi, M. and Gordon, A.D., Reasoning about Cryptographic Protocols in the Spi Calculus. In *CONCUR '97: Concurrency Theory*, Warzaw, Poland. Lecture Notes in Computer Science, Vol. 1243. Springer, pp. 59–73, 1997.

[171] Hoare, C.A.R., Communicating Sequential Processes. *Communications of the ACM* **21**(8), 666–677, 1978.

[172] Hoare, C.A.R., *Communicating Sequential Processes*. Prentice Hall, 1985.

[173] Ryan, P. and Schneider, S. (Eds.), *The Modelling and Analysis of Security Protocols: The CSP Approach*. Addison-Wesley, 2001.

[174] Lowe, G., Casper: A Compiler for the Analysis of Security Protocols. In *Proceedings of the 10th Computer Security Foundations Workshop (CSFW '97)*. IEEE Computer Society, pp. 18–30, 1997.

[175] Meadows, C.A., The NRL Protocol Analyzer: An Overview. *Journal of Logic Programming* **26**(2), 113–131, 1996.

[176] Paulson, L.C., Proving Properties of Security Protocols by Induction, pp. 70–83, 1997.

[177] Otway, D. and Rees, O., Efficient and Timely Mutual Authentication. *ACM SIGOPS Operating Systems Review* **21**(1), 8–10, 1987.

[178] Clarke, E.M., Jha, S. and Marrero, W., Verifying security protocols with Brutus. *ACM Transactions on Software Engineering and Methodology (TOSEM)* **9**(4), 443–487, 2000.

[179] Aiello, L.C. and Massacci, F., Verifying Security Protocols as Planning in Logic Programming. *ACM Transactions on Computational Logic (TOCL)* **2**(4), 542–580, 2001.

[180] Boreale, M. and Buscemi, M.G., Experimenting with STA, A Tool for Automatic Analysis of Security Protocols. In *Proceedings of the 2002 ACM symposium on Applied Computing*, Madrid, Spain. ACM Press, pp. 281–285, 2002.

[181] Ruys, T.C., Towards Effective Model Checking. PhD Thesis, Universiteit Twente, Enschede, The Netherlands, 03 2001.

[182] Jøsang, A., Security protocol verification using SPIN. In J-Ch. Grégoire (Ed.), *Proceedings of the First SPIN Workshop*. INRS-Télécommunications, Montréal, Quebec, Canada, 10 1995.

[183] Maggi, P. and Sisto, R., Using SPIN to Verify Security Properties of Cryptographic Protocols. In D. Bonaki and S. Leue (Eds.), *Proceedings of the 9th International SPIN Workshop*, Grenoble, France, April 11–13, 2002, Lecture Notes in Computer Science, Vol. 2318. Springer, pp. 187–204, 2002.

[184] Wagner, D., Pushdown Model Checking for Security. In P. Godefroid (Ed.), *Proceedings of the 12th International SPIN Workshop*, San Francisco, CA, USA, August 22–24, 2005, Lecture Notes in Computer Science, Vol. 3639. Springer, p. 1, 2005.

[185] Khan, A.S., Mukund, M. and Suresh, S.P., Generic Verification of Security Protocols. In P. Godefroid (Ed.), *Proceedings of the 12th International SPIN Workshop*, San Francisco, CA, USA, August 22–24, 2005, Lecture Notes in Computer Science, Vol. 3639. Springer, pp. 221–235, 2005.

[186] Rothmaier, G., Kneiphoff, T. and Krumm, H., Using SPIN and Eclipse for Optimized High-Level Modeling and Analysis of Computer Network Attack Models. In P. Godefroid (Ed.), *Proceedings of the 12th International SPIN Workshop*, San Francisco, CA, USA, August 22–24, 2005, Lecture Notes in Computer Science, Vol. 3639, Springer, pp. 236–250, 2005.

[187] Mercer, E. and Jones, M., Model Checking Machine Code with the GNU Debugger. In P. Godefroid (Ed.), *Proceedings of the 12th International SPIN Workshop*, San Francisco, CA, USA, August 22–24, 2005, Lecture Notes in Computer Science, Vol. 3639. Springer, pp. 251–261, 2005.

[188] AVISPA Team. *Automated Validation of Internet Security Protocols and Applications (AVISPA); HLPSL Tutorial; A Beginner's Guide to Modelling and Analysing Internet Security Protocols*. AVISPA IST-2001-39252, http://www.avispa-project.org, 1.1 edition, 06 2006.

[189] Mödersheim, S., Models and Methods for the Automated Analysis of Security Protocols. PhD Thesis, ETH Zürich, Information Security Group, Haldeneggsteig 4, CH-8092 Zürich, 2007.

[190] Turuani, M., The CL-Atse Protocol Analyser. In *Proceedings of the 17th International Conference on Term Rewriting and Applications, RTA 2006*, Seattle, WA, USA, August 12–14, 2006, Lecture Notes in Computer Science, Vol. 4098, Springer, pp. 277–286, 2006.

[191] Armando, A. and Compagna, L., SATMC: A SAT-Based Model Checker for Security Protocols. In *Proceeding of the 9th European Conference on Logics in Artificial Intelligence, JELIA 2004*, Lisbon, Portugal, September 27–30, 2004, Lecture Notes in Computer Science, Vol. 3229. Springer, pp. 730–733, 2004.

[192] Compagna, L., SAT-Based Model-Checking of Security Protocols. PhD Thesis, Dipartimento di Informatica, Sistemistica e Telematica, Università degli Studi di Genova, Italy, 08 2005.

[193] Boichut, Y., Kosmatov, N. and Vigneron, L., Validation of Prouvé protocols using the automatic tool TA4SP. In *Proceedings of the 3rd Taiwanese-French Conference on Information Technology*, Nancy, France, pp. 467–480, 2006.

[194] Glouche, Y. and Genet, T., *SPAN – A Security Protocol ANimator for AVISPA; User Manual*. INRIA/IRISA, LANDE Project, http://www.irisa.fr/lande/genet/span/, 1.1 edition, 02 2007.

[195] Boichut, Y., Genet, T., Glouche, Y. and Heen, O., Using Animation to Improve Formal Specifications of Security Protocols. In *The 2nd Joint Conference on Security in Network Architectures and Information Systems SAR-SSI'07*, Annecy-France, June 12–15, 2007, 06 2007.

[196] Glouche, Y., Genet, T., Heen, O. and Courtay, O., A Security Protocol Animator Tool for AVISPA. In *ARTIST2 Workshop on Security Specification and Verification of Embedded Systems*, Pisa, Italy, 05 2006.

[197] Pancho, S., Paradigm Shifts in Protocol Analysis. In *Proceedings of the 1999 Workshop on New Security Paradigms*, Caledon Hills, Ontario, Canada. ACM Press, pp. 70–78, 1999.

[198] Clark, J.A. and Jacob, J.L., Protocols Are Programs Too: The Meta-heuristic Search for Security Protocols. *Information and Software Technology* (**43**), 891–904, 2001.

Index

About the Author

The author has been working with cellular systems and security both in industry and in academia. He started out working for LM Ericsson in Norway, testing NMT system software (NMT was an 1G cellular system), as well as system testing ISDN exchanges before leaving Ericsson.

Later he worked as a consultant for System Sikkerhet AS, a company which carried out security evaluation of system software for communictaions systems for the Nowegian Military. This work focused on system evaulation according to an extended Orange Book set of criteria. In 1995 he joind Telenor, the Norwegian incumbent telecom operator. Here he first developed and taught advanced system courses in cellular systems and cellular signalling protocols. In 1998 he joined the Telenor R&D department, where he worked with system security and mobile/cellular systems, including UMTS and LTE. From 1999 to 2009 he has been the Telenor delagate to the 3GPP SA3 (security) work group.

The author holds a BSc.Hons in Computing Science from the University of Newcastle upon Tyne (UK), a MSc in information technology from the Norwegian University of Science and Technology (NTNU) (Trondheim, Norway) and a PhD in security/mobile communication from Aalborg University (Denmark).

Currently, the author is affilliated with Agder University (Norway) (adjunct associate Professor) and runs a consultant business.

RIVER PUBLISHERS SERIES IN STANDARDISATION

Volume 1
Wireless Independent Living for a Greying Population
Lara Srivastava
2009
ISBN: 978-87-92329-22-6